DECODING THE GENOMIC CONTROL OF IMMUNE REACTIONS

The Novartis Foundation is an international scientific and educational charity (UK Registered Charity No. 313574). Known until September 1997 as the Ciba Foundation, it was established in 1947 by the CIBA company of Basle, which merged with Sandoz in 1996, to form Novartis. The Foundation operates independently in London under English trust law. It was formally opened on 22 June 1949.

The Foundation promotes the study and general knowledge of science and in particular encourages international co-operation in scientific research. To this end, it organizes internationally acclaimed meetings (typically eight symposia and allied open meetings and 15–20 discussion meetings each year) and publishes eight books per year featuring the presented papers and discussions from the symposia. Although primarily an operational rather than a grant-making foundation, it awards bursaries to young scientists to attend the symposia and afterwards work with one of the other participants.

The Foundation's headquarters at 41 Portland Place, London W1B 1BN, provide library facilities, open to graduates in science and allied disciplines. Media relations are fostered by regular press conferences and by articles prepared by the Foundation's Science Writer in Residence. The Foundation offers accommodation and meeting facilities to visiting scientists and their societies.

Information on all Foundation activities can be found at http://www.novartisfound.org.uk

The Australian Phenomics Facility (APF) is a leading organization that provides a unique approach to the study of physiology and health via the elucidation of the genome–phenome code. Many of the health problems we face today—obesity, cancer, autoimmune diseases, allergy, cardiovascular disease—stem from a discordance between genetics and environment. The APF has pioneered technologies to produce libraries of mice with informative gene alterations, and ways of using those libraries for new genome–phenome discoveries.

The APF is located in Canberra, Australia. Information on all APF activities can be found at http://www.apf.edu.au.

Novartis Foundation Symposium 281

DECODING THE GENOMIC CONTROL OF IMMUNE REACTIONS

John Wiley & Sons, Ltd

Published in 2007 by John Wiley & Sons Ltd,
The Atrium, Southern Gate,
Chichester PO19 8SQ, UK

National 01243 779777
International (+44) 1243 779777
e-mail (for orders and customer service enquiries): cs-books@wiley.co.uk
Visit our Home Page on http://www.wileyeurope.com
or http://www.wiley.com

Other Wiley Editorial Offices

John Wiley & Sons Inc., 111 River Street, Hoboken, NJ 07030, USA

Jossey-Bass, 989 Market Street, San Francisco, CA 94103-1741, USA

Wiley-VCH Verlag GmbH, Boschstr. 12, D-69469 Weinheim, Germany

John Wiley & Sons Australia Ltd, 33 Park Road, Milton, Queensland 4064, Australia

John Wiley & Sons (Asia) Pte Ltd, 2 Clementi Loop #02-01, Jin Xing Distripark, Singapore 129809

John Wiley & Sons Canada Ltd, 6045 Freemont Blvd, Mississauga, Ontario, Canada L5R 4J3

Wiley also publishes its books in a variety of electronic formats. Some content that appears in print may not be available in electronic books.

Novartis Foundation Symposium 281
x + 209 pages, 21 figures, 5 tables

Anniversary Logo Design: Richard J. Pacifico

British Library Cataloguing in Publication Data

A catalogue record for this book is available from the British Library

ISBN 978-0-470-02755-4

Typeset in 10½ on 12½ pt Garamond by SNP Best-set Typesetter Ltd., Hong Kong
Printed and bound in Great Britain by T. J. International Ltd, Padstow, Cornwall.
This book is printed on acid-free paper responsibly manufactured from sustainable forestry, in which at least two trees are planted for each one used for paper production.

Contents

Novartis Foundation Symposium on Decoding the genomic control of immune reactions, held at University House, The Australian National University, Canberra, 7–9 March 2006, in collaboration with the Australian Phenomics Facility and The Australian National University

Editors: Gregory Bock (Organizer) and Jamie Goode

This symposium was based on a proposal made by Chris Goodnow

Participants

Warren S. Alexander The Walter and Eliza Hall Institute of Medical Research, and Cancer & Haematology Division, 1G Royal Parade, Parkville, 3050 Victoria, Australia

Gabrielle Belz The Walter and Eliza Hall Institute, 1G Royal Parade, Parkville, 3050 Victoria, Australia

Warwick J. Britton Centenary Institute of Cancer Medicine and Cell Biology, and Department of Medicine, Central Clinical School, University of Sydney, Bld DO6, Sydney, NSW 2006, Australia

Jean-Laurent Casanova Laboratoire de Génétique Humaine des Maladies Infectieuses, Université de Paris René Descartes-INSERM U550, Faculté de Médecine Necker, 156 rue de Vaugirard, F-75015 Paris, France

Matthew Cook Australian National University Medical School, Department of Immunology, The Canberra Hospital, Canberra, Australia

Jason Cyster Howard Hughes Medical Institute and Department of Microbiology and Immunology, University of California, San Francisco, San Francisco, CA 94143-0414, USA

Simon Foote The Menzies Research Institute, 17 Liverpool Street, Hobart 7000, Tasmania, Australia

Anny Fortin Emerillon Therapeutics Inc., 416 de Maisonneuve West, Suite 1000, Montreal, H3A 1L2, Canada

Christopher C. Goodnow *(Chair)* John Curtin School of Medical Research, The Australian National University, Canberra, ACT 2061, Australia

Philippe Gros Department of Biochemistry, Center for the Study of Host Resistance, and Cancer Center, McGill University, Montreal, PQ, Canada H3G 1Y6

David Hume Institute for Molecular Biosciences, The University of Queensland, Brisbane, QLD 4072, Australia

Gunasegaran Karupiah Immunology and Genetics Division, John Curtin School of Medical Research, Building 54, Australian National University, Canberra, ACT 2000, Australia

Jim Kaufman Institute for Animal Health, Compton, Newbury, Berkshire RG20 7NN, UK

Kong-Peng Lam Laboratory for Immunology, Centre for Molecular Medicine, 20 Biopolis Way, #08-01 Centros, Singapore 138668

Charles Mackay Garvan Institute of Medical Research, 384 Victoria St, Darlinghurst, NSW 2010, Australia

Rick M. Maizels Institute of Immunology and Infection Research, University of Edinburgh, West Mains Road, Edinburgh EH9 3JT, UK

Bernard Malissen Centre d'Immunologie de Marseille-Luminy, INSERM-CNRS, Parc Scientifique et Technologique de Luminy—Case 906, F-13009 Marseille, France

Grant Morahan Western Australia Institute for Medical Research, Perth Campus, Level 6, MRF Building, Rear, 50 Murray Street, Perth, WA 6000, Australia

Anthony Scalzo Centre for Experimental Immunology, Lions Eye Institute, 2 Verdun St, Nedlands, WA 6009, Australia

Christian Schönbach Nanyang Technological University, School of Biological Sciences, Division of Genomics and Genetics, 60 Nanyang Drive, Singapore 637551

Steven Stacker Angiogenesis Laboratory, Ludwig Institute for Cancer Research, Post Office Royal Melbourne Hospital, Parkville, Victoria 3050, Australia

Stephen Turner Department of Microbiology and Immunology, The University of Melbourne, Parkville, Victoria 3010, Australia

Carola Vinuesa John Curtin School of Medical Research, The Australian National University, Canberra, ACT 2061, Australia

Edward K. Wakeland Center for Immunology, University of Texas Southwestern Medical Center, 5323 Harry Hines Blvd, Dallas, TX 75235, USA

Douglas Young Centre for Molecular Microbiology and Infection, Division of Investigative Sciences, Flowers Building, Room 3.42, South Kensington Campus, Imperial College London, London SW7 2AZ, UK

Chair's introduction

Chris Goodnow

John Curtin School of Medical Research, The Australian National University, Canberra, ACT 2061, Australia

There are lots of meetings in immunology—too many, probably—where we discuss all the latest hot topics. There are also lots of meetings on genomics. But what attracted me about this meeting is the unique format, with an equal balance between people presenting work and then the discussions that follow. This allows us to address some of the bigger issues. This is particularly important at this time in immunology. Rather than focusing on all the details, we have an opportunity to step back and try to identify the strategic gaps in the field.

We know the genome sequence is a tremendous toolkit. It is a parts list of the whole immune system. But we are increasingly becoming aware that it isn't entirely clear which way is best for using this genome sequence to conquer some of the big issues in immunology that have been there for a long time. If we are going to achieve something in this meeting that will be different from any other kind of meeting, it will be to tease out some of the issues about what does work. What sorts of approaches are working for different people here? What kinds of conceptual issues are on the boil? Where are the conceptual gaps and technical gaps? These are the big strategic issues that the unique combination of people here will be well placed to addressed. I hope our discussions will tease out some of these issues.

Some of the issues that might be illuminated by the papers are as follows. What sequences in the genome encode different kinds of response, such as resistance to infection, memory, tolerance and immunopathology? There are issues about reductionist approaches, and the need to focus on individual genes to get at mechanism, versus the problem of trying to understand the system as a whole. What is 'systems biology', and how can we use this concept? There are issues of bottom–up approaches, starting with gene sequences and trying to work up to whole immune systems or organs, versus top–down approaches, of starting with populations of humans or animals and working down to mechanisms underlying variation in immune responses. I look forward to a good meeting.

Transcriptional regulatory networks in macrophages

David A. Hume*, Christine A. Wells† and Timothy Ravasi‡

**ARC Special Research Centre for Functional and Applied Genomics and CRC for Chronic Inflammatory Diseases, Institute for Molecular Bioscience, University of Queensland, Q4072, Australia, † School of Biomolecular and Biomedical Science, (The Eskitis Institute for Cell and Molecular Therapies) Griffith University, Nathan Campus Kessels Rd, Q4111, Australia and ‡ Department of Bioengineering, Jacobs School of Engineering, University of California, San Diego. 9500 Gilman Drive, La Jolla, CA 92093-0412, USA*

Abstract. The functions of macrophages in the innate immune system require the constitutive expression of a wide range of myeloid-specific genes, including various pattern recognition receptors, as well as the inducible expression of a suite of genes required to initiate inflammation and eliminate pathogens. Our overall aim is to understand the transcriptional networks that underlie both macrophage-specific transcription and the response to pathogen components such as lipopolysaccharide (LPS). The approaches used include detailed functional analysis of specific promoters, such as that of the CSF1 receptor, global cDNA microarray expression profiling, high throughput real-time PCR analysis of all the transcription factors encoded by the mammalian genome, full length cDNA library construction and sequencing, CAGE analysis to identify specific promoters used in macrophages and motif analysis to detect candidate *cis*-acting elements in co-regulated genes in macrophages. This review discusses some of the progress in moving towards a transcriptional network model for mouse macrophage activation by LPS, as well as insight into the role of alternative promoter usage and polyadenylation in generating functional protein variants that impact on signalling in macrophages.

2007 Decoding the genomic control of immune reactions. Wiley, Chichester (Novartis Foundation Symposium 281) p 2–24

The mononuclear phagocyte system

The mononuclear phagocyte system (MPS) is defined as a family of cells comprising bone marrow progenitors, blood monocytes and tissue macrophages. Macrophages are a major cell population in most of the tissues in the body, and their numbers increase further in inflammation, wounding and malignancy (Hume 2006, Hume et al 2002). Through their endocytic and cytotoxic activities, they provide a first line of defence against pathogens in the innate immune system. They contribute to antigen processing and presentation, and in turn their effector functions are activated in response to products of the acquired immune system, notably antibodies (via Fc receptors), and T cell products (e.g. interferon γ and interleukin 4). Apart from their roles in immunity, macrophages also contribute to many aspects of homeostasis, vascularization, normal development, tumour progression and wound healing in part through their adaptation to recognize and remove cells undergoing apoptosis (Henson & Hume 2006, Lichanska & Hume 2000) and through their secretion of growth factors and proteolytic enzymes (Hume 2006, Hume et al 2002).

Mononuclear phagocytes as they appear in tissues share a number of features (Hume 2006, Hume et al 2002):

- Stellate morphology and ultrastructural evidence of endocytic activity observed by light and electron microscopy.
- Expression of histochemical markers such as non-specific esterase, lysosomal hydrolases and ecto-enzymes.
- Evidence of endocytic activity.
- The presence of cell surface proteins (such as F4/80, CD14 and CSF1R) defined by monoclonal antibodies.

A database of images of tissue macrophages detected using the F4/80 monoclonal antibody in all mouse organs can be found at *www.macrophages.com*. The precise anatomical location of these cells supports a physiological role for tissue macrophages in the development, structure and homeostasis of organs. Not all macrophages express the F4/80 marker. To find an alternative we have produced the MacGreen transgenic mouse lines, in which macrophage-restricted expression of an EGFP transgene is directed by the promoter of the CSF1R (c-fms) gene. The proliferation, differentiation and survival of macrophages is controlled by macrophage colony-stimulating factor (CSF-1), and the receptor is expressed in a myeloid-restricted manner. These mice provide a resource for many aspects of macrophage biology, including the isolation of macrophage populations from specific organs (MacDonald et al 2005, Sasmono et al 2003). A library of images showing the location of macrophages detected using the Fms-EGFP transgene is also available at *www.macrophages.com*.

Our ability to monitor this population of cells in real time and *in vivo* raises the interesting question of exactly how their numbers are monitored, controlled and replenished; the cellular 'systems biology' of the mononuclear phagocyte system. At least in the normal steady state, the availability of CSF-1 appears to be regulated via macrophage-dependent endocytosis and destruction, which provides a simple link between macrophage numbers and the production of cells by the marrow. Accordingly, the level of CSF-1 rises substantially in mice that lack the CSF-1 receptor (Dai et al 2002). The control is undoubtedly more complex than that; since CSF-1 expression is regulated in many different tissues (Sweet & Hume 2003) and at least some macrophage populations may be controlled by local proliferation (Hume 2006). Furthermore, several other growth factors can regulate macrophage production from bone marrow (Hume 2006).

Mammalian cells are complex systems. Their functions require the appropriate interactions of millions of individual components in appropriate order in time and space. Although biological scientists commonly study cells as if they were static entities, individual cells never exist in a steady state. Like an entire multicellular organism, each individual cell is born by cell division, adapts to its environment to carry out a particular function that is determined by its genetically programmed response to that environment, and then dies. If all the components of a biological system interacted with each other, the complexity would be impenetrable. However, as in most complex systems, individual components within a cell interact and organize in defined modules that we recognize as structures, pathways and regulatory frameworks. The traditional approach to understanding cell biology has been to identify those individual modules, dissect their components, and then attempt to determine how they can be reassembled. Systems Biology can be considered an alternative approach, in which one actually measures all the components and their interactions within a biological system, and then assembles them into modules (Aderem 2005). The advantage of this approach is that it makes no assumptions; the disadvantage is that it is constrained by the scale of information that can be obtained.

To understand the function of a cell in perfect detail, we would need to know the identity of every macromolecule and metabolite, its location, abundance and chemical status across the lifetime of each cell. Since cells in a population influence each other's behaviour in many ways, we would also need to know the same information about its neighbours; every other cell in the body! This is clearly not possible. However, the problem can simplified if one takes the view that the entire control framework of a mammalian cell is encoded in its genome, and exerted through the selective production of RNA, the process of transcription. If we know the identity of every transcript produced by the genome and the way that transcription changes with time, we can infer the fundamental control pathways. Certainly, different tissues express distinct transcriptional profiles—transcriptional lineage markers—that can be used to identify sub-populations of cells in complex tissues, for example the stem cell populations profiled in the Stem Cell Gene Anatomy Project (see *www.scgap.org*; Challen et al 2005, 2004). The process of lineage differentiation is arguably the process of restricting the transcriptional network of a cell to those genes that define its function. Ultimately, the life of that cell can be viewed by a transcriptional network.

Transcription is itself controlled by RNA; the precise way this feedback control operates may be very complex. RNA may act directly on the induction, processing or stability of another transcript. Non-coding RNAs are a major, regulated, output of the mammalian genome (Katayama et al 2005, Ravasi et al 2006). Alternatively, the RNA may encode a protein. That protein may participate directly in transcription control in the nucleus, or it may be a secreted protein that directly or indirectly initiates a signalling cascade that ultimately regulates transcription. Hence, the control system of biological systems is ultimately a network of linked transcriptional switches. The completion of multiple eukaryote genome sequences and the rapid progress towards definition of the complete transcriptional output (the transcriptome) has given us access to all of the components of the network. The capacity to analyse the complete transcriptional profile of cells, and the advent of technologies that allow us to capture protein/RNA and DNA interactions within this transcriptional context offers us the tools to identify the connectivities within the network, and ultimately, the control system.

There are many reasons why macrophages are an ideal cell type in which to apply a systems approach to transcription control in a mammalian context:

(1) They can be obtained as reasonably homogeneous primary cell cultures; from peripheral blood monocytes in humans, or by cultivation of bone marrow cells or peritoneal lavage in laboratory animals.
(2) There are cell lines available in both mouse and human that replicate many of the characteristics of primary macrophages and which can be transiently or stably transfected.
(3) The range of mammalian biologies/pathways that can be studied using macrophages is substantial; they can be regarded in significant measure as the archetypal cell.
(4) They alter their function in distinct ways in response to many different extracellular signals acting through multiple distinct signalling pathways, and most importantly these changes in the macrophage regulatory networks can be measured over time (Sester et al 2005, Rehli et al 2005, Schroder et al 2004, Wells et al 2003a).
(5) There are substantial natural, or introduced, genetic influences on the function of macrophages (Wells et al 2003a, Beutler 2005, Beutler et al 2005, Fortier et al 2005). These provide perturbations of the network which can be analysed to establish connections between nodes.
(6) Since macrophages are both our primary line of defence against pathogens and mediators of much of the pathology of infectious, inflammatory and malignant disease, a fundamental understanding of the way their function is controlled in likely to translate into rational development of human therapies.
(7) Their differentiation from progenitor cells, which involves stable epigenetic changes, can also be studied in detail *in vitro* (Tagoh et al 2002).

Macrophage activation and deactivation

The life of a macrophage 'family' starts with the division of a pluripotent progenitor, and subsequent division of successive progeny associated with progressive transcription of genes that are required for the mature function of macrophages, and which distinguish the macrophage from other cell types. Individual cells receive further signals that direct them to adhere to endothelial cells, transmigrate, and adapt to whatever environment they encounter. If that environment is a site of inflammation, rather than a place of normal macrophage residence, the cell will be 'activated' to eliminate the challenge, be it dead cells or pathogens. Subsequently, the cell must be deactivated and either die or migrate out of the site to permit resolution and restoration of normal tissue architecture (or in some cases, a scar). This life story requires the macrophage to respond to hundreds, perhaps thousands of distinct signals produced by other cells, including other macrophages, as well as pathogen products. Because macrophages are able to cause damage to normal tissue, their arsenals must be unleashed only when required to deal with a pathogen or other emergency. This requires a very rigorous mechanism for self/non-self discrimination. This distinction is mediated in large measure by so-called pattern recognition receptors. The toll-like receptors (TLRs), which recognize cell wall components and nucleic acid molecules that are conserved amongst pathogen classes, are the archetypal pattern recognition receptors, but many other classes of receptors contribute to selective pattern recognition.

Of all the pathogen components, lipopolysaccharide (LPS, also known as endotoxin), a structural constituent of Gram-negative bacterial cell walls, has been studied in most depth. LPS action on macrophages requires the TLR4 receptor, and involves multiple signalling pathways leading to regulated transcription of thousands of genes. The system has been so widely studied that a survey of the literature alone can generate a list of hundreds of target genes discovered one at a time (Hume et al 2002). In considering the response of macrophages to LPS, it is important to identify a number of features that we feel are generic to all transcriptional networks and which provides challenges to design and interpretation:

(1) The initiation response to LPS at a single cell level is, to some extent at least, all-or-nothing (Sester et al 1999, Sweet & Hume 1996). It is not clear whether individual signalling pathways linked to TLR4 have distinct thresholds, nor how a presumably analogue signal (extent of receptor occupancy) is translated into a digital outcome.

(2) At some level, each gene has its own threshold for transcriptional activation. This can be seen in the observation that individual genes have LPS dose response curves that differ by orders of magnitude (Costelloe et al 1999).

(3) Individual genes also have a degree of all-or-nothing transcriptional activity at the single cell level, determined in part by stable epigenetic modifications and in part by chance (Hume 2000).
(4) No signal is received in isolation. Other signals alter the dose response curve, magnitude and nature of the LPS response. Three of particular importance are the growth factor, CSF1 (Sester et al 2005), the major T cell product interferon γ (Schroder et al 2004), and glucocorticoids, which are commonly used therapeutically to ablate macrophage activation and inflammation (Ogawa et al 2005).
(5) The response of cells to any signal is influenced by individual genetic variation in the species (Wells et al 2003a). In the case of macrophage response to LPS and other pathogen-associated molecules, this difference is particularly prevalent because the nature of the macrophage response is under strong selection pressure from pathogens.
(6) The response is a sequential cascade of transcriptional regulatory events. Individual genes (notably the proinflammatory cytokines) are induced transiently and then repressed (Wells et al 2003a). The transient nature of the transcriptional regulatory response to LPS (and any other pathogen challenge) is absolutely crucial to resolution.

The last point highlights an interesting feature of the control architecture of the macrophage transcriptional network. We have reviewed elsewhere the observation that many of the most LPS-inducible transcripts are actually not effectors of immunity, but feedback controllers that act at every level of the signalling pathway from the receptor to transcription, as well as repressors of the effectors (such as protease inhibitors and antiinflammatory cytokines) (Wells et al 2005). One might consider that this multilevel control exists to ensure robustness, but in fact mutation in any individual control leads to disordered or excessive macrophage activation. A prediction of the observation that the feedback control is itself inducible, and the system is sensitive to that feedback control, is that it will be relatively insensitive to inhibitors that act on the afferent pathways (Wells et al 2005).

Transcriptome data sets and the macrophage transcriptional network

Macrophages as a cell population provide one of the most complex sources of transcripts in any cell type. In fact, we have performed comparative transcriptional profiling of LPS-stimulated mouse macrophages compared to 17 day embryos and embryonic stem cells (Fig. 1). The macrophage and embryonic stem (ES) cell populations in combination cover the large majority of the protein-coding transcripts on a complex microarray platform. The complexity of macrophages as a source of transcripts was also exploited by the FANTOM consortium

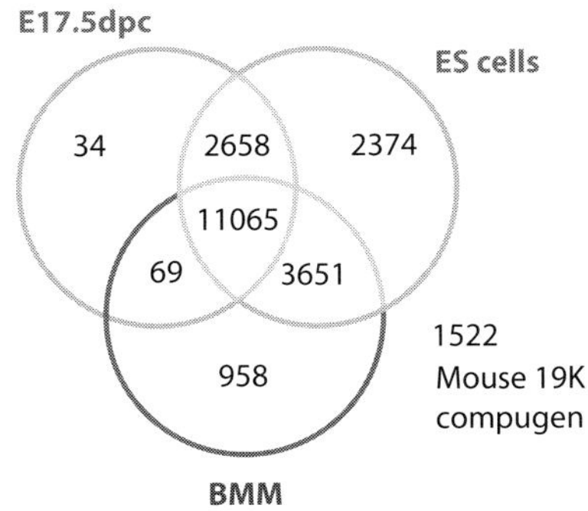

FIG. 1. A Venn diagram showing the overlap in transcription profiles between macrophages (BMM), embryonic tissues (E17.5dpc) and embryonic stem (ES) cells. RNA was prepared from bone marrow-derived macrophages grown in CSF-1 (BMM), a pool of 17 day mouse embryos, and embryonic stem cells, each from 129/Sv mice, and compared on the Mouse Compugen microarray platform. The diagram shows the numbers of genes expressed above the significant detection threshold. 1522 were not detected in any sample. Note that macrophages and ES cells each express genes that are unique to that cell population, and in combination, express the large majority of genes on the chip.

in its efforts to identify the full diversity of transcripts encoded by the mouse genome (Wells et al 2003b). Arising from that effort, we have recently analysed the impact of alternative initiation, splicing and polyadenylation on a selected part on the macrophage proteome, focusing specifically on signalling molecules that contribute to the LPS response (Wells et al 2006). This analysis built upon the original finding that LPS acts upon an internal promoter in the IRAK-2 signalling molecule to generate inducible expression of a natural feedback repressor, a component of control network mentioned above. Using splicing arrays, we confirmed the expression of many additional examples of variant forms of signalling molecules that could form part of the control architecture.

Table 1 contains a summary of the microarray studies on macrophages that are currently available in the public domain. Our own analysis has focused on a particular mouse cellular system. Bone marrow-derived macrophages cultured in CSF-1 are exposed to LPS across a time course from 1 h to 18–24 h. Our interest in this system is partly based upon an interest in interactions between the CSF-1 and LPS signalling pathways (Sester et al 1999, 2005, Sweet & Hume 2003). The system is interesting because LPS, in common with agonists of other TLRs, blocks CSF-1-induced proliferation, but substitutes for CSF-1 in maintaining survival (Sester et al 1999, 2005). Aside from extensive array profiling on a number of different platforms (Nilsson et al 2006, Wells et al 2006, 2003a, *symatlas.gnf.org*), and in a number of different mouse strains which produce radically different transcription profiles

TABLE 1 Summary of microarray studies on macrophages currently available in the public domain

Description	*Database*	*Species*	*Platform*	*Reference*
Time course of LPS-stimulated bone morrow derived macrophages	*www.macrophages.com*	Mouse	RIKEN 20K cDNA microarray	Wells et al 2003 (PMID: 12826024)
Bone morrow derived macrophages stimulated for 16h with interleukin (IL)4	*www.macrophages.com*	Mouse	NIA 15K mouse cDNA microarrays	Rehli et al 2005 (PMID: 15908341)
Time course of LPS-stimulated bone morrow derived macrophages	*www.macrophages.com*	Mouse	RIKEN 60K cDNA microarray	Nillson et al 2006 (PMID: 16698233)
GPX-Macrophage Expression Atlas (range of macrophage cell types)	*www.gti.ed.ac.uk/GPX*	Human and Mouse	Mix platforms	Grimes et al 2005 (PMID: 16343346)
Time course of LPS-stimulated RAW264.7	*www.ncbi.nlm.nih.gov/geo*	Mouse	IMB custom cDNA arrays	Ravasi et al 2002 PMID: 11751944
Time course of LPS-stimulated bone morrow derived macrophages	*http://fantom3.gsc.riken.jp*	Mouse	CAGE libraries	Kawaji et al 2006 (PMID: 16381948)
GM-CSF matured human blood monocytes with or without LPS or glucocorticoid	*http://linkage.garvan.unsw.edu.au/public/microarrays/Arthritis_Inflammation/index.html*	Human	Affymetrix GeneChip technology	N/A
HL-60 myeloid cell line, PMA maturation to macrophages.	*http://www.broad.mit.edu/cgi-bin/cancer/datasets.cgi*	Human	Affymetrix GeneChip technology	Tamayo, et al 1999 (PMID: 10077610

TABLE 1 *(continued)*

Description	*Database*	*Species*	*Platform*	*Reference*
Systems Biology and Medicine Database (range of macrophage cell types)	*www.lsbm.org*	Human	Affymetrix GeneChip technology	N/A
SymAtlas of the Genome Institute of the Novartis Research Foundation	*http://symatlas.gnf.org/SymAtlas*	Human and Mouse	Affymetrix GeneChip technology	Su et al 2004 (PMID: 15075390)
An atlas of human gene expression (adult monocytes)	*http://mpss.licr.org*	Human	MPSS	Jongeneel et al 2005 (PMID: 15998913)
Gene expression Omnibus (GEO) (range of macrophage cell types)	*www.ncbi.nlm.nih.gov/geo*	Human and Mouse	Mix platforms	Wheeler et al 2006 (PMID: 16381840)
Time course of *Salmonella typhimurium* induced gene expression on RAW264.7 macrophage cell line	*http://cmdr.ubc.ca/salmonellaarray*	Mouse	Atlas mouse cDNA arrays	Rosenberger et al 2000 (PMID: 10820271)
Human monocytes exposed to a range of pathogens	*http://web.wi.mit.edu/young/pathogens*	Human	Affymetrix GeneChip technology x	Nau et al 2002 (PMID: 11805289)

(Wells et al 2003a), we have also used this system in the context of analysis of the promoters of the mouse genome.

The transcriptional output of any cell is influenced by the interaction between nuclear proteins and DNA sequences in the vicinity of the gene (promoters, enhancers, repressors). The identification of promoters is dependent upon accurate determination of the transcription start site. Recently, this has become possible on a genome-wide scale using cap analysis of gene expression (CAGE), essentially high throughput 5′ RACE/SAGE, technology (Carninci et al 2006). This combined with sequences of full length cDNAs and other polling technologies (diTags) has enabled the first genome-wide view of transcription initiation. To expedite access to this information, we have created a portal at *www.macrophages.com/bioinfoweb*. For the analysis of macrophages, CAGE sequencing has been performed in multiple independent libraries across the time course of activation by CSF-1 and LPS, providing an index of relative promoter use. Such data highlight the fact that a large proportion of the protein-coding genes in the genome have multiple promoters expressed in different cell types and tissues, or, as in the case of IRAK2, regulated differently in the same cell (Carninci et al 2005, 2006). Figures 2 and 3 show examples of two key macrophage-expressed transcription factors, STAT1 and NF-κB (p105). In each case, the presence of alternative promoters allows the locus to generate alternative protein isoforms that may have quite distinct biological activities.

The CAGE data provide us with a survey of the set of transcription start sites (TSS) utilized specifically in macrophages. Importantly, CAGE tags also provide a measure of frequency of promoter usage–information lacking from normalized EST or FL-cDNA datasets. The sequences flanking these start sites can be extracted and analysed as a class for the presence of over-represented sequence motifs for broad classes of transcription factors. Not surprisingly, given the known architecture of TATA-less myeloid promoters (Rehli et al 1999, Ross et al 1998) such an analysis reveals over-representation of purine-rich motifs recognized by the Ets transcription factor family amongst constitutive macrophage-specific genes, as well sites for the LPS-inducible transcription factor NF-κB among inducible genes (Carninci et al 2006). A more penetrating analysis can be done using either experimentally-validated binding sites for known transcription factors (e.g. Transfac, Jaspar) or *a priori* motif detection (Nilsson et al 2006). The final piece in the puzzle is the identity of the transcription factors that are actually present and likely to bind the candidate motifs in the promoters.

The FANTOM project has also provided a comprehensive set of candidate transcriptional regulators encoded by the mouse genome. Transcription factors (TF) were amongst the most highly alternatively spliced class in the F3 dataset—many new variants of known TF have been discovered (Ravasi et al 2003, Kummerfeld & Teichmann 2006). Aside from known transcription factors, additional candidates

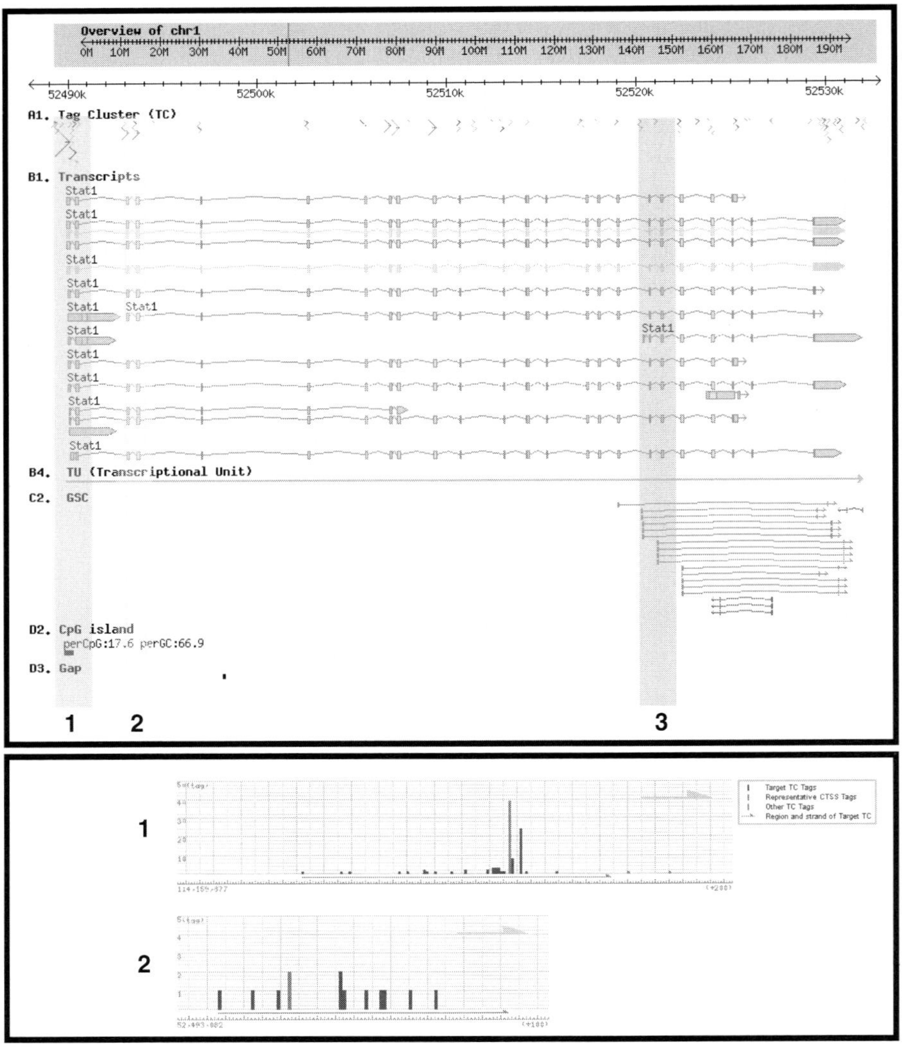

FIG. 2. Screenshot of the mouse Stat1 gene structure from the Macrophages BioinfoWeb (*www.macrophages.com/bioinfoweb*). Top panel shows the chromosome 1 region with: (A1) Mapping of all the CAGE tags defining the CAGE transcription starting sites (CTSS). (B2) The full-length cDNAs mapping of Stat1. (B4) Boundaries of the defined transcriptional unit for Stat1. (C2) GSC diTags mapping, see text for description. The inferred Stat1 transcriptional starting sites, generating three main isoforms are highlighted by the shaded vertical lines. Bottom panel shown the histogram distribution of the CAGE tags from several tissues for Stat1 isoform 1 and 2.

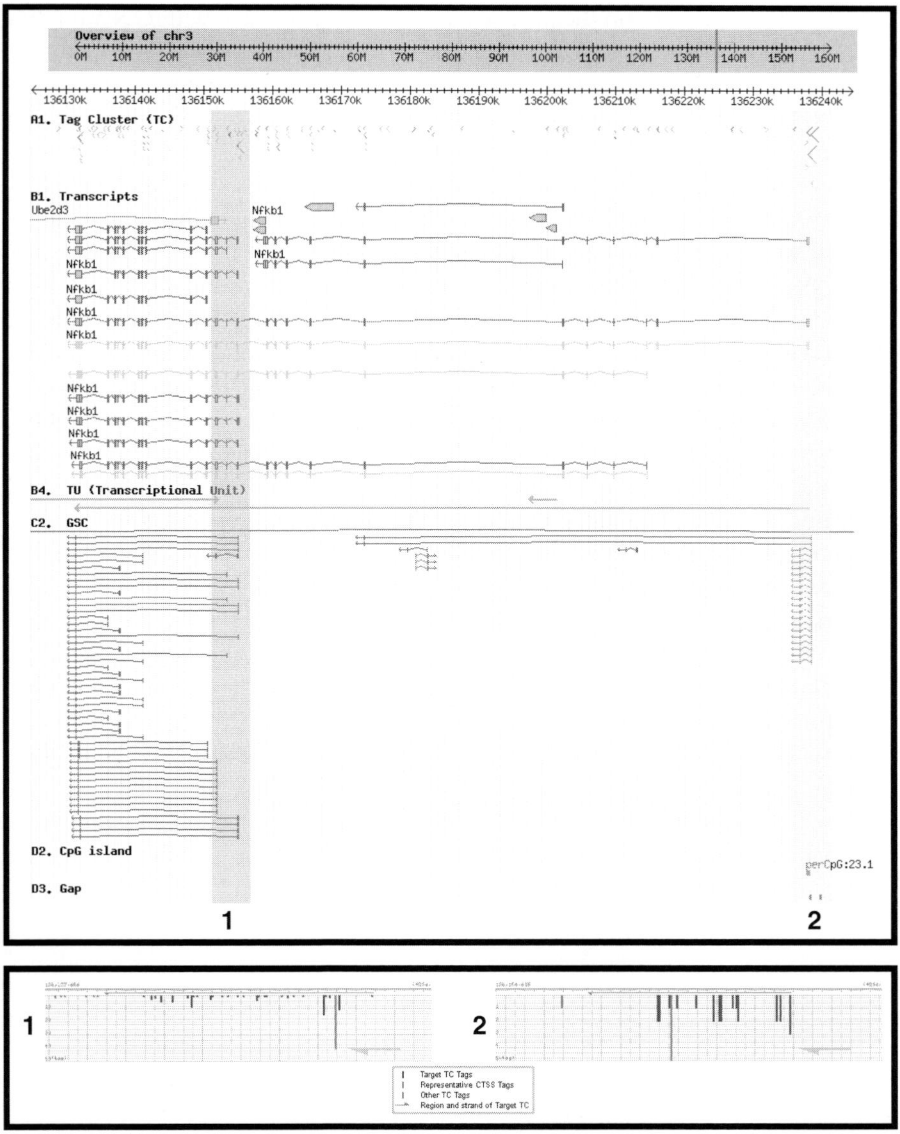

FIG. 3. Screenshot of the mouse NF-κB1 (p105) gene structure from the Macrophages BioinfoWeb (*www.macrophages.com/bioinfoweb*). Top panel shows the chromosome 3 region with: (A1) Mapping of all the CAGE tags defining the CAGE transcription starting sites (CTSS). (B2) The full-length cDNAs mapping of NF-κB1 (p105). (B4) Boundaries of the defined transcriptional unit for NF-κB1. (C2) GSC diTags mapping, see text for description. The inferred NF-Kb1 (p105) transcriptional starting sites, generating 2 main isoforms, are highlighted by shaded vertical lines. Bottom panel shown the histogram distribution of the CAGE tags from several tissues for NF-κB1 (p105) isoform 1 and 2.

were identified based upon the presence of domains associated with DNA binding or other nuclear functions. The set of known functional transcription factors in macrophages has been reviewed elsewhere (Himes & Hume 2003), but was anticipated to be far from comprehensive. qRT-PCR analysis as carried out to identify the complete set of transcripts encoding nuclear proteins that is expressed by macrophages across the time course of response to LPS (Nilsson et al 2006). The increased sensitivity of qRT-PCR is important, because many transcription factors are expressed at relatively low levels, around the detection thresholds for microarrays. An interesting finding in this global analysis is that around 60% (850/1439) of all the transcripts encoding nuclear proteins were detectable in macrophages (Nilsson et al 2006). The notable exclusions were members of gene families known to be involved in developmental processes wherein they are expressed transiently in tightly defined embryonic locations. They include most members of the Hox, Sox, Gata, Neurog, Lhx and Tbx families. This observation alone suggests that other nuclear proteins that are not expressed in macrophages will probably also have functions in development that require absolutely stringent control.

Taking all of the pieces of information together, it is possible to construct a first-pass transcriptional control network for response of macrophages to LPS. The analysis links transcription factor expression from qRT-PCR, the promoter architectures (i.e. presence/absence of transcription factor binding sites) of all the transcripts expressed in macrophages (the macrophage transcriptome) and semantic approaches to infer regulatory circuits (Nilsson et al 2006). The predicted network has the characteristics of a scale-free network; with a relatively small number of nodes that are highly connected. Amongst these predicted nodes are most of the known players in LPS responses: the Ets family, NF-κB and interferon-response element binding proteins. The unexpected finding was an implied role for the stress response pathway and the transcription factors Nrf2, Atf2 and Atf3 (Nilsson et al 2006).

Where do we go from here?

The stringency and quality of predicted transcriptional networks in macrophages (or any other system) can be increased by acquiring and integrating additional pieces of information:

(1) Parallel analysis of a second species to identify conserved modules. This is the focus of the next phase of the FANTOM project, which is directed towards human. There is already a great deal of CAGE, transcript and expression profiling information for human macrophages in the public domain.
(2) Literature data. Software packages such as Cytoscape (*www.cytoscape.org*); Cell Designer (which has been used to create a network model for macrophage

signalling (Oda et al 2004), and Cell Illustrator (*www.gene-networks.com*) have been used to store, interrogate and visualize signalling networks. Connectivities can also be confirmed using text mining tools such the Agilent Literature Search software *www.agilent.com* and iHOP (*www.ihop-net.org/UniPub/iHOP*).

(3) Protein data. Proteomic analysis of mammalian systems is in its infancy and transcription factors are at the limits of sensitivity for current technologies (Vigneault & Guerin 2005). However, chromatin immunoprecipitation, coupled to arrays (so-called ChIP on Chip) can be used to confirm the expression of transcription factors that are considered major nodes in a network, and that they are bound to the promoters of their predicted target genes at the times when they are inferred to act. High throughput approaches to assessing candidate protein–protein interactions (PPIs) (Suzuki et al 2001), will also inform the construction of networks and large scale PPI measurements are starting to become available for mammalian systems (Peri et al 2004, Rual et al 2005).

(4) Epigenetic studies. We would like to know which candidate factor binding sites on particular genes are occupied, and how the DNA and associated histone and non-histone proteins change with time. Many studies correlate transcription factor binding site occupancy/promoter usage with chromatin remodelling (e.g. the CSF1R locus is dynamically remodelled during macrophage differentiation (Tagoh et al 2002), but relatively few have achieved the precise level of detail necessary to unravel the control mechanisms in time and space (Metivier et al 2006).

(5) Manipulation of the network. Ultimately, a useful network model makes predictions about the likely consequences of varying the functional expression of specific transcription factors. In mice, many transcription factors implicated in macrophage differentiation and activation have been deleted in the mouse germline, but the analysis of the phenotypic consequences in terms of transcription control have been limited. Of course, knockout of a gene such as PU.1 which is essential for macrophage production provides limited opportunity for analysis (Lichanska et al 1999). In some cases, the results are unexpected. For example, the macrophage-specific factor TFE-C appears to regulate a rather small subset of genes involved in response to interleukin 4 (Rehli et al 2005). The alternative approach is to create hypomorphic mutations in the germ line (Rosenbauer et al 2004), knock it down using siRNA, or to over-express the transcription factor of interest in transgenic animals or cells. The most informative experiments would permit the change in the network to be monitored with time.

Many transcription factors act in tandem; for example our own work has highlighted interactions between PU.1 and MITF, and Ets and Ap1 in macrophage differentiation (Fowles et al 1998, Luchin et al 2001). Finally, many transcription factors (notably NF-κB) are controlled at least partly by post-translational modifications, and by regulated translocation between the nucleus and cytoplasm. Information about these aspects requires traditional cell biology and biochemistry approaches which are currently not readily extended to high throughput for mammalian systems. A network model can prioritize particular transcription factors for such detailed study.

Summary

The pre-genomic era was characterized in some measure by a race to acquire information. The post-genomic era saw a secondary race to identify the transcriptional outputs of mammalian genomes, and the impact of proteomic, and ultimately a metabolomic, revolution is probably still to come. However, at this point in time we do have many of the tools needed to create predictive network models of a cellular system like macrophage activation and transcription control.

References

Aderem A 2005 Systems biology: its practice and challenges. Cell 121:511–513

Beutler B 2005 The Toll-like receptors: analysis by forward genetic methods. Immunogenetics 57:385–392

Beutler B, Georgel P, Rutschmann S et al 2005 Genetic analysis of innate resistance to mouse cytomegalovirus (MCMV). Brief Funct Genomic Proteomic 4:203–213

Carninci P, Kasukawa T, Katayama S et al 2005 The transcriptional landscape of the mammalian genome. Science 309:1559–1563

Carninci P, Sandelin TA, Lenhard B et al 2006 Genome-wide analysis of mammalian promoter architecture and evolution. Nat Genet 38:626–635

Challen GA, Martinez G, Davis MJ et al 2004 Identifying the molecular phenotype of renal progenitor cells. J Am Soc Nephrol 15:2344–2357

Challen G, Gardiner B, Caruana G et al 2005 Temporal and spatial transcriptional programs in murine kidney development. Physiol Genomics 23:159–171

Costelloe EO, Stacey KJ, Antalis TM, Hume DA 1999 Regulation of the plasminogen activator inhibitor-2 (PAI-2) gene in murine macrophages. Demonstration of a novel pattern of responsiveness to bacterial endotoxin. J Leukoc Biol 66:172–182

Dai XM, Ryan GR, Hapel AJ et al 2002 Targeted disruption of the mouse colony-stimulating factor 1 receptor gene results in osteopetrosis, mononuclear phagocyte deficiency, increased primitive progenitor cell frequencies, and reproductive defects. Blood 99:111–120

Fortier A, Min-Oo G, Forbes J, Lam-Yuk-Tseung S, Gros P 2005 Single gene effects in mouse models of host: pathogen interactions. J Leukoc Biol 77:868–877

Fowles LF, Martin ML, Nelsen L et al 1998 Persistent activation of mitogen-activated protein kinases p42 and p44 and ets-2 phosphorylation in response to colony-stimulating factor 1/c-fms signaling. Mol Cell Biol 18:5148–5156

Henson PM, Hume DA 2006 Apoptotic cell removal in development and tissue homeostasis. Trends Immunol 27:244–250

Himes SR, Hume DA 2003 Transcription factors that regulate macrophage development. Handbook Exp Pharmacol 158:11–40

Hume DA 2000 Probability in transcriptional regulation and its implications for leukocyte differentiation and inducible gene expression. Blood 96:2323–2328

Hume DA 2006 The mononuclear phagocyte system. Curr Opin Immunol 18:49–53

Hume DA, Ross IL, Himes SR et al 2002 The mononuclear phagocyte system revisited. J Leukoc Biol 72:621–627

Katayama S, Tomaru Y, Kasukawa T et al 2005 Antisense transcription in the mammalian transcriptome. Science 309:1564–1566

Kummerfeld SK, Teichmann SA 2006 Alternative splicing and expression of a mouse transcription factor repertoire. PLoS Genetics, in press

Lichanska AM, Browne CM, Henkel GW et al 1999 Differentiation of the mononuclear phagocyte system during mouse embryogenesis: the role of transcription factor PU.1. Blood 94:127–138

Lichanska AM, Hume DA 2000 Origins and functions of phagocytes in the embryo. Exp Hematol 28:601–611

Luchin A, Suchting S, Merson T et al 2001 Genetic and physical interactions between Microphthalmia transcription factor and PU.1 are necessary for osteoclast gene expression and differentiation. J Biol Chem 276:36703–36710

MacDonald KP, Rowe V, Bofinger HM et al 2005 The colony-stimulating factor 1 receptor is expressed on dendritic cells during differentiation and regulates their expansion. J Immunol 175:1399–1405

Metivier R, Reid G, Gannon F 2006 Transcription in four dimensions: nuclear receptor-directed initiation of gene expression. EMBO Rep 7:161–167

Nilsson R, Bajic V, Katayama S et al 2006 Transcriptional network dynamics in macrophage activation. Genomics 88:133–142

Oda K, Kimura T, Matsuoka Y et al 2004 Molecular interaction map of a macrophage. AfCS Research Reports 2:1–12

Ogawa S, Lozach J, Benner C et al 2005 Molecular determinants of crosstalk between nuclear receptors and toll-like receptors. Cell 122:707–721

Peri S, Navarro JD, Kristiansen TZ et al 2004 Human protein reference database as a discovery resource for proteomics. Nucleic Acids Res 32:D497–501

Ravasi T, Huber T, Zavolan M et al 2003 Systematic characterization of the zinc-finger-containing proteins in the mouse transcriptome. Genome Res 13:1430–1442

Ravasi T, Suzuki H, Pang KC et al 2006 Experimental validation of the regulated expression of large numbers of non-coding RNAs from the mouse genome. Genome Res 16:11–19

Rehli M, Lichanska A, Cassady AI, Ostrowski MC, Hume DA 1999 TFEC is a macrophage-restricted member of the microphthalmia-TFE subfamily of basic helix-loop-helix leucine zipper transcription factors. J Immunol 162:1559–1565

Rehli M, Sulzbacher S, Pape S et al 2005 Transcription factor Tfec contributes to the IL-4-inducible expression of a small group of genes in mouse macrophages including the granulocyte colony-stimulating factor receptor. J Immunol 174:7111–7122

Rosenbauer F, Wagner K, Kutok JL et al 2004 Acute myeloid leukemia induced by graded reduction of a lineage-specific transcription factor, PU.1. Nat Genet 36:624–630

Ross IL, Yue X, Ostrowski MC, Hume DA 1998 Interaction between PU.1 and another Ets family transcription factor promotes macrophage-specific Basal transcription initiation. J Biol Chem 273:6662–6669

Rual JF, Venkatesan K, Hao T et al 2005 Towards a proteome-scale map of the human protein-protein interaction network. Nature 437:1173–1178

Sasmono RT, Oceandy D, Pollard JW et al 2003 A macrophage colony-stimulating factor receptor-green fluorescent protein transgene is expressed throughout the mononuclear phagocyte system of the mouse. Blood 101:1155–1563

Schroder K, Hertzog PJ, Ravasi T, Hume DA 2004 Interferon-gamma: an overview of signals, mechanisms and functions. J Leukoc Biol 75:163–189
Sester DP, Beasley SJ, Sweet MJ et al 1999 Bacterial/CpG DNA down-modulates colony stimulating factor-1 receptor surface expression on murine bone marrow-derived macrophages with concomitant growth arrest and factor-independent survival. J Immunol 163:6541–6550
Sester DP, Trieu A, Brion K et al 2005 LPS regulates a set of genes in primary murine macrophages by antagonising CSF-1 action. Immunobiology 210:97–107
Suzuki H, Fukunishi Y, Kagawa I et al 2001 Protein-protein interaction panel using mouse full-length cDNAs. Genome Res 11:1758–1765
Sweet MJ, Hume DA 1996 Endotoxin signal transduction in macrophages. J Leukoc Biol 60:8–26
Sweet MJ, Hume DA 2003 CSF-1 as a regulator of macrophage activation and immune responses. Arch Immunol Ther Exp (Warsz) 51:169–177
Tagoh H, Himes R, Clarke D et al 2002 Transcription factor complex formation and chromatin fine structure alterations at the murine c-fms (CSF-1 receptor) locus during maturation of myeloid precursor cells. Genes Dev 16:1721–1737
Vigneault F, Guerin SL 2005 Regulation of gene expression: probing DNA-protein interactions in vivo and in vitro. Expert Rev Proteomics 2:705–718
Wells CA, Ravasi T, Faulkner GJ et al 2003a Genetic control of the innate immune response. BMC Immunol 4:5
Wells CA, Ravasi T, Sultana R et al 2003b Continued discovery of transcriptional units expressed in cells of the mouse mononuclear phagocyte lineage. Genome Res 13:1360–1365
Wells CA, Ravasi T, Hume DA 2005 Inflammation suppressor genes: please switch out all the lights. J Leukoc Biol 78:9–13
Wells CA, Chalk A, Forrest A et al 2006 Alternative transcription of the toll-like receptor signalling cascade. Genome Biol 7:R10

DISCUSSION

Goodnow: One of the issues I keep encountering with big microarray data sets is the following. You mentioned that half of the transcripts are transcription factors. I struggle with the issue of how well annotated genes are in terms of their ontologies. What is your sense of how good or bad the state of ontologies in the immune system is?

Hume: The problem of annotation is an ongoing one. It was one of the major challenges of the RIKEN project. In the most recent effort, the RIKEN project participants came to the conclusion that a computational approach was superior to a human curation. Individuals made idiosyncratic decisions about what they would call a specific gene and how much weight they would give to the apparent presence of a DNA-binding domain. Annotation comes down to setting fixed rules and letting machines do it. Human curation has been a major problem in an annotation, rather than an asset. If you look at what we have done with quantitative real time for both mouse and human, the set that we took to look at was some 2000 genes which includes anything that has an annotated nucleotide binding domain. This potentially includes zinc fingers that are not involved in transcription at all. In that set there are ubiquitin-associated processes where the proteins have zinc fingers. This set covers every transcription factor that I know of, so it is erring on the side of inclusion rather than exclusion.

Wakeland: When you were describing these broad transcription start point promoters, you mentioned that they were more flexible in terms of evolution of quantitation. Have you looked at these promoter regions in some of the genes that vary among different strains of mice, in terms of their expression in macrophage differentiation, to show a change?

Hume: We haven't done that. But we have looked at the difference in activity between mouse and human promoters in mouse and human cells. For example, the CSF-1 receptor promoter has sequence variants across species. I think the promoters of innate immune-associated genes are evolving much more rapidly than other promoters in the genome, because they are under such strong evolutionary selection. We find that the mouse c-Fms promoter is much more active in mouse cells than the human CFMS promoter. We can actually assay whether that is the case due to mutations in key proximal promoter elements and also in the start site itself. There are base pair shifts between mouse and human. The obvious thing to do now is to look across mouse strains, to search for promoter polymorphisms. Now that we have them anchored and have access to genomes of more than one mouse strain, we could trawl for them in the genome sequence and then look for the most divergent ones. There is one that is quite well known and which probably underlies what we are looking at: the interferon (IFN)β promoter is polymorphic in both mouse and human. This probably has a significant effect on the gene expression profile differences between mouse strains, since interferon is a key endogenous regulator.

Vinuesa: When one does comparative microarray analysis across strains, how confident can we be that the differences observed are genuine expression differences, rather than just polymorphisms between the different mouse strains?

Hume: If you look at an individual gene that is radically different in its expression between two mouse strains, it doesn't mean that this gene itself is genetically different in any way. It could be a pleiotropic effect. This is inherent in what I said: if we cross C57BL and BALB/c, we get a phenotype that is identical to C57BL in the F1. The most likely explanation for that is that there is a single dominant variation in C57BL/6 compared with BALB/c that controls the whole LPS signalling cascade.

Vinuesa: Is there a systematic study analysing how many oligos actually bind on the polymorphic areas in the different models?

Hume: We have confirmed this with three different platforms, but if it is a detection difference, those are actually full length cDNA arrays, so variants don't influence detection. The differences between the strains have also been confirmed with both Affymetrix and Compugen oligo arrays.

Lam: You mentioned that the different inbred mouse strains respond differently to lipopolysaccharide (LPS) stimulation. How relevant, therefore, are the mouse studies to humans? And wouldn't it be better to use the outbred rather than inbred mouse strains?

Hume: I gave a talk at a TLR (Toll-like receptor) meeting a few years ago, where my first slide was, 'Is the C57BL/6 mouse a good model for a mouse?', let alone a human. That is a legitimate question. Inbred mouse strains have tremendous utility in helping us to understand the events that occur in immune response in every other kind of biology that we want to study, because they reduce the number of variables, but other than a few select populations around the world, there are no seriously inbred humans. If you look at the LPS-inducible profile, it does differ among individual humans, but nothing like the extent it does in mice.

Scalzo: You touched on the off-regulation of genes and how this is important in controlling unwanted damage. Can you elucidate a bit further on the use of microarrays to get the mechanisms?

Hume: A couple of years ago I published a review on what we called 'inflammation suppressor' genes (Wells et al 2005). The interesting thing about the control architecture of the LPS response in macrophages is that some 30 or 40 of the most inducible genes are actually inhibitors of the response. These include MAP kinase phosphatase, tristetraprolin, and inhibitors of IRAK. If you look at all the splice variants we have recently described, it seems that there is a negative regulator of every single step of the Toll signalling cascade, from outside the cell through to transcription, and also all of the effectors. The most inducible genes are antiproteases. This invites the question: why are there so many inducible feedback controllers? You might say that it has to be robust. But this is not true. Mutation of any one of them causes the process to go out of control. This is an interesting intellectual question: why is it like that? I think it has to be like that to enable evolution to act on it. It has to be the case that very small genetic changes in any one of the feedback controllers changes the outcome of the inflammation response. Every individual in this room has a subtly different innate immune response to any particular pathogen. Extending this further, genetic selection acts on the feedback controllers, not on the afferent pathways. In other words, it acts on the brakes rather than the accelerators.

Karupia: In your studies with the LPS stimulation the macrophages from BL/6 and DBA, you found that the responses were very similar.

Hume: There are significant differences; they are just similar to each other. The interesting thing in those data sets that we want to come back to is that there are probably hundreds of null mutations on each of those inbred backgrounds. There are transcripts which we simply don't detect at all in one strain that are easily detectable in all the other strains. We don't detect them at any time point, so it is not just that one array didn't work. This suggests we could save ourselves a lot of knockouts if we were to go back with high stringency and trawl through the existing mouse genetic reservoir. Look for anything you are interested in, establish a repository of expression profiles in all the mouse strains, and you may find that a particular mouse strain doesn't express the gene.

Cyster: You talked about inflammatory hotspots in the genome. What sort of numbers of genes do they contain? Is this a conserved feature between mouse and human?

Hume: I think they had to have at least 10 genes. This is conserved between mouse and human in that those regions are syntenic in human.

Wakeland: How large were those regions? Wasn't the most populated one on chromosome 7 in the mouse?

Hume: I don't have the information in my head. That mapping work was mainly by Chris Wells.

Morahan: What is the functional significance of those multiple independent transcription sites you described?

Hume: One example that we have is UDP glucuronyl transferase, subunit 1A, which has seven different transcription start sites, each of which encodes a separate ATG. There are seven different alternative first exons, each of which would give a different N-terminus. The example I showed of gelsolin is the difference between plasma gelsolin and cytoplasmic gelsolin. These alternative forms are produced by using alternate promoters. In the case of IRAK, there is a constitutive promoter and an internal inducible promoter that generates a natural dominant negative feedback control of that gene. We don't understand the control architecture unless we know the full set of promoters that can drive each of the genes. They make different proteins and they are differently regulated.

Morahan: Is this true for most of these promoters, that they direct transcription of mRNAs that encode different proteins?

Hume: For the broad promoters, I think this is just the way that they work. We are talking about broad promoters versus truly independent promoters. There is a bit of a grey area between. If you take a typical CpG island, certain tissues will use one region of it and other tissues will use another region. We arbitrarily decide whether we call this one promoter or two. If you go within them, what is the significance of using multiple start sites? I don't know; I think it is just the way it works. It doesn't involve TATA binding protein binding to one site with the initiator 30 bp downstream. It is a binding recognition that we think involves EWS and FUS/TLS, which binds to anything that has four out of five Gs in a row, and starts somewhere between 20 and 30 bp downstream of that sequence. Because this is so loose, you get multiple independent start sites.

Goodnow: What is impressive about these kinds of approaches is that if you have a specific gene, and you want to know how it is regulated across all these different conditions, things like what you described with the sym atlas is tremendously powerful for that bottom–up gene-driven approach. But what I often see as a real gap for us is if you are coming at this problem top–down. Say you are Jean-Laurent Casanova and you have a family who are getting recurrent infections. You have mapped the susceptibility to a region on a particular chromosome. Then there is a huge gap: you would like to know what genes are involved in macrophage regulation in that interval. Coming at it from this top–down approach, is there a way to capture this wealth of information from the arrays in a form that someone coming from the top–down can instantly identify the interesting genes?

Hume: That is the architecture that we want to put together next: we would like to have a way of automating links. These exist, but things like Ensembl are impenetrable unless you spend your life looking at them. It will be great when we can get genome maps and predicted transcripts linked straight off to relevant expression data. The trouble is that there is just so much of it. If you go with your favourite gene into a repository like GEO on NCBI, or ArrayExpress at EBI, you'll spend the rest of the day trawling through entries to find something that is actually useful. I suppose it comes down to establishing your own resource for your own purpose, and setting up that architecture locally, so that you only go after the information that you particularly want.

Goodnow: Is there a way of creating some kind of a language that would extract the different genes that have the different expression patterns? Could you create some kind of tag, so if you are interested in the genes in macrophages that have this tag, for example responding to LPS but not TLR7, you could extract that information?

Hume: That is relatively easy. Just as an indication of something that we did, the NOD2 gene is a susceptibility gene for Crohn's disease. The gene is expressed specifically in monocytes and is down-regulated as they become macrophages, and it is inducible by LPS. We took that classification and went searching for other genes in the transcriptome that have this regulatory profile on the assumption that other susceptibility genes that have the same phenotypic outcome would have the same regulation. This is a bit of a stretch, but all of the known Crohn's disease loci have that pattern, and one of the genes that we came out with was the human orthologue of F4/80 (EMR1) which is a candidate as a Crohn's disease susceptibility locus. You can use intuition and say that if we know a gene that has a phenotype and want to find other genes that are candidates to cause the same phenotype, the chances are they will be regulated in roughly the same way. So you just go into the cluster that has your lead gene, and you trawl within that cluster at various levels of stringency and association. This identifies candidates which can be mapped to the genome to see whether they are in regions of interest.

Goodnow: Are there public domain tools for this? If you wanted to find all the genes that are regulated by TLR4, for example, could you pull them out?

Hume: You can do this on the Symatlas site (*symatlas.gnf.org*). The example I showed was cathepsin K. We used this to identify some 300 osteoclast-specific genes that are strongly validated. The list that you get contains all 30 or so well-known osteoclast-specific genes. Charles Mackay has this kind of data on his website as well. You can take an anchor gene, and if want to find everything that is coexpressed with IL6 in human leucocytes, you can go in and do a fine correlate, and download the data locally. That isn't particularly challenging.

Mackay: There are more sophisticated approaches that have been done, although I am not too familiar with them. We had a collaborator who did this nearest neighbour analysis. This assesses all the genes that are expressed in a similar pattern. That is quite powerful. These sorts of analyses will become more available.

Hume: If you do it in both mouse and human you can merge the common data sets to identify genes with common regulatory patterns across the species. For example, I trawled for PU.1, my favourite transcription factor, and looked for all the genes across every tissue that are coexpressed with PU.1 at varying levels of stringency. Then we looked for PU.1 sites in the promoters of those genes. Every one of them is a candidate PU.1 target based on both promoter analysis and coexpression. You don't have to be all that sophisticated to start to walk into this field.

Stacker: How do you integrate the proteomic data with your genomic data? When there are proteases working actively *in vivo*, how do you integrate these two pieces of information, to come up with a weighting as to which one is more important?

Hume: The proteome is the next frontier. We are trying to publish a paper comparing the transcriptome of neutrophils and macrophages. These are highly purified populations, and these data are supported by human data. Neutrophils and macrophages are almost indistinguishable at the transcriptome level. There is a small number of neutrophil-specific transcripts and almost no macrophage-specific transcripts. Yet the proteomes are massively different. There are all these transcripts expressed by neutrophils but not transcribed. For my favourite macrophage-specific gene, the transcript is abundant in neutrophils. There is no protein at all: transcription doesn't equal translation. We are only really just scraping the surface of the tools that we need to get into the proteome. In my lab we are investing in getting after the plasma membrane proteome and the nuclear proteome, but it is a morass. We are using segments of DNA to purify every protein that can associate with them. We are also using protein–protein interaction data, using one protein to pull down every other protein, looking for functional relationships.

Stacker: Is that where we really need to go for target identification? If you want to investigate a pathway and block it in a human situation, do you need to understand the proteome in order to find the true targets?

Hume: We need to understand the control architecture: how it works. I mentioned about the feedback control being more important than the afferents. The drug industry has spent a lot of time trying to make p38 inhibitors. p38 is one of the on signals, but it itself induces the off signals. If you damp down p38 the best you can achieve is to delay the inflammatory process, but you won't switch it off unless you completely ablate p38. Not only do you not switch it on, but also you don't switch on the off signal. Paradoxically, p38 inhibitors are proinflammatory because the off signal fails in the absence of a strong enough on signal. It is complex. If we understand the control systems we will be able to make rational guesses as to which molecules are likely to be good drug targets, and are likely to switch the pathway off. I would be looking for allosteric activators of the off signal here.

Stacker: In the angiogenesis area there is a big push to find natural promoters of the inhibitors.

Hume: The best indicator we have is with the glucocorticoids, which induce most of the off signals. This is our most effective drug: look for the genes that the glucocorticoids induce: these will be your natural feedback controls.

Reference

Wells CA, Ravasi T, Hume DA 2005 Inflammation suppressor genes—please turn out all the lights. J Leukocyte Biol 78:9–13

The RIKEN mouse transcriptome: lessons learned and implications for the regulation of immune reactions

Christian Schönbach[1]

Immunoinformatics Research Team, Advanced Genome Information Technology Group, RIKEN Genomic Sciences Center (GSC), 1-7-22 Suehiro-cho, Tsurumi, Yokohama 230-0045, Japan

Abstract. Notably, the technology and analysis methods of the RIKEN mouse full-length cDNA project have contributed a lot to the capture of the transcriptional output of the mouse genome and the description of its combinatorial nature. However, one corollary of this large-scale transcript resource is the dichotomy of vast and missing information. As such, the transcriptional and translational output of yet unknown size following non-canonical principles remains to be established and interpreted. The importance of identifying immune-related transcripts and establishing their molecular functions in context of complex immune system diseases is clear: knowledge about the transcriptome can advance the understanding of immune system regulation. Deciphering the logic of transcriptomes is critical for understanding the ontogeny and effector functions of immune cells, but it is not sufficient. The next challenge will lie in the combined sampling and integrated analysis of genomic elements, transcripts, proteins and metabolites.

2007 Decoding the genomic control of immune reactions. Wiley, Chichester (Novartis Foundation Symposium 281) p 25–37

DNA microarrays and gene chips have been successfully applied to identify signatures of genes expressed during immune cell differentiation or responses (Chtanova et al 2005). Yet, the information on gene expression patterns is limited by the coverage and quality of the probe sets which depend on *a priori* knowledge of transcripts and thus the completeness of a cell's transcriptome. Towards this end, the RIKEN Mouse Encyclopedia (*http://genome.rtc.riken.go.jp*) full-length cDNA project in conjunction with the FANTOM (Functional Annotation of Mouse/Mammalians; *http://fantom.gsc.riken.jp*) consortium enormously increased

[1]Present address: Division of Genomics and Genetics, School of Biological Sciences, Nanyang Technical University, 60 Nanyang Drive, Singapore 637551

the coverage of the transcriptome and knowledge about its variation and regulation (Hayashizaki 2003). The isolation and sequencing of full-length cDNAs from 204 C57/BL6 tissues and cells and 33 different mouse strains (i.e. NOD) or cell lines (i.e. DBA/2 L5178Y-R melanoma cells) resulted in 102801 sequences (Carninci et al 2005), representing 64.7% of all *Mus musculus* sequences (158810) available at the time of GenBank release 139. The sequences, excluding T cell receptors and immunoglobulins, were clustered into 44147 transcriptional units (TUs) and 45142 transcriptional frameworks (TKs).

Both TKs and TUs are operational definitions of transcripts that cluster in one orientation to the same genomic regions, with TKs delineating fusions of unrelated TUs by requiring shared transcription start sites or splicing events. The definitions may appear to be arbitrary, but they were essential for developing an annotation pipeline and for analysing alternative splicing, transcription start sites and sense–antisense transcripts. Five web-accessible databases integrate human-curated functional annotations of all cDNA sequences with information on genome mapping, transcription start/end tags, CpG islands, promoter elements, gene ontology and protein motifs. Aside from the usefulness of the full-length cDNAs as publicly available resources or tools for improving the mouse genome annotation and DNA microarrays, the real value lies in the associated functional information and technology platforms for identifying the patterns and signals that regulate the output of the genome. An important prerequisite in this quest is the knowledge of the transcript boundaries along with transcript and promoter expression information for each 'gene'.

An open system for gene regulatory analyses

Cap Analysis Gene Expression (CAGE) (Shiraki et al 2003) and Gene Identification Signature (GIS) (Ng et al 2005) are two new technologies based on sequencing and genome mapping of DNA tags corresponding to the 5′ ends (CAGE) or 5′ and 3′ ends (GIS) of transcripts. In contrast to microarrays, both tagging approaches resemble an open system permissible for the discovery of new transcript boundaries. In the FANTOM3 study DNA tagging revealed more than 181000 transcript boundaries (Carninci et al 2005). The frequency of CAGE tags within a TU and a tag cluster provide the expression information on 'gene' and promoter levels, respectively. In addition, 7.1 million mapped mouse CAGE tag data for 145 libraries were complemented with 3.1 million human CAGE full-length cDNA data for 24 libraries of the H-Inv consortium (*http://www.h-invitational.jp*). Although the two data sets are not integrated on the database viewer level the raw data facilitate analyses of variation in transcriptional start sites, gene expression, promoter properties and usage between mouse and human in general (Bajic et al 2006), as well as across different tissue cells of gene families such as mammalian antimicrobial peptides (Brahmachary et al 2006). Since the sensitivity of CAGE depends only on the number of sequenced tags, rare or even single transcripts can be captured.

MicroRNAs and other non-protein-coding RNAs

The CAGE technology enabled the detection of many weakly expressed non-protein-coding RNAs (ncRNAs) and tissue cell-specific transcription initiated at the stem-loop of micro RNAs (miRNAs). Thus far, miRNAs derived from introns of protein-coding pre-mRNAs have been known only as sequence-specific regulators on a post-transcriptional level. For example, miR-375 suppresses glucose-induced insulin secretion via its target mRNA myotrophin (Poy et al 2004). The findings reported by the FANTOM3 consortium together with recent computational estimates of 200 miRNAs, on average, per target mRNA (Krek et al 2005) imply a critical role of miRNAs in gene regulatory networks. Further investigations along the lines of Ramkissoon's (2006) work which demonstrated different expression patterns of haematopoietic cell-specific miRNAs in normal and malignant cell lineages should yield new insights into the regulation of cancer and autoimmunity.

The majority of known ncRNAs represent genuine polyadenylated transcripts rather than transcriptional noise. Together with a still-small fraction of non-polyadenylated ncRNAs they make up over 50% (22 839) of all TUs. A growing body of literature supports the regulatory roles of ncRNA in imprinting (i.e. *Air*), inducible processes by external stimuli, tissue cell differentiation and development. Possible mechanisms include long-range interaction with the promoter region of a neighbouring gene, concordant and to a lesser extent reciprocal sense–antisense expression (Katayama et al 2005). Until now, a few ncRNAs have been analysed with respect to their role in immune response regulation. Non-coding transcript in T cells (NTT) is one of the ncRNAs that is specifically expressed in activated T cells and induced by interferon (INF)γ (Amarante et al 2005). However, understanding the full extent of ncRNAs in immune response regulation will require not only a systematic interrogation on gene expression level (Ravasi et al 2006) but also new approaches for integrating these results with assay data on protein and protein signalling pathway levels.

Alternative splicing and protein functions

Alternative transcription start sites which can induce the splicing of cryptic initial exons and alternative splicing play an important role in expanding the protein-coding capacity of the genome. In FANTOM3 about 32% (16 274) of the 51 135 protein-coding sequences were reported as novel. Of these 76% (12 303) represented previously unknown splice variants of known TKs, reflecting the increase in tissue cell sources and depth of sequencing. In total, 40% (12932) of protein-coding TKs contain splice variants that may indirectly affect protein functions by changing the expression level (variation in 5′ or 3′ UTR) or directly by altering the coding potential and motif/domain content.

The variety of probably inducible and cell-specific transcripts suggests the presence of sequence patterns in addition to the known splice enhancer/silencers motifs that recruit splicing factors. For example, the motif TGAA is frequently found in constitutive exons but rarely in cryptic exons (Zavolan et al 2006). Conversely, AGGG and TTGACA are over-represented in cryptic exons and introns preceding non-first cryptic exons, respectively (Bock and Schönbach, unpublished results 2005). Besides, there appears to be some dependency between splicing, transcription start site, polyadenylation site and exon length. Using genomic exon coordinates of GenBank 141 mRNA sequences mapped to mouse genome assembly (mm4) we found that initial exons are on average 120 bp longer than middle exons and significantly shorter last exons or single exons. Alternatively spliced exons are, on average, 250 bp longer than constitutively spliced exons. Mining the patterns associated with variant and constitutive exons will be important for understanding the positional requirements and effects of signals that facilitate the differential expression of transcript variants in immune response regulation and effector functions.

The TU encoding the transcriptional regulator Aire, which plays an important role in establishing self-tolerance, contains 17 transcripts representing six variants produced by alternative splicing and alternative transcription initiation. The alternative transcription start site in one embryonic stem cell-derived variant (GenBank accession AK131938) results in an N-terminal truncated isoform lacking the SAND domain which is likely to affect the transcription of genes controlled by Aire. Since the SAND domain is necessary for nuclear targeting, the Aire isoform may not enter the nucleus to interact with the TTATTA-box of its target promoters (Purohit et al 2005).

Another Aire variant (AK153690) derived from the thymus of 3 day neonate mice retains an intron between exon 10 and 11 which is supported by two thymus and mammary gland-derived expressed sequence tags (ESTs). Since the retained intron is unlikely an artifact it begs the question why does a cell contains transcripts that may result in nonsense-mediated decay (NDM)? Data coming from studies of mutually exclusive splice variants of rat fibroblast growth factor Fgfr2 (Jones et al 2001, Lewis et al 2003) suggest a functional linkage of alternative splicing and NDM to regulate tissue or developmental stage-specific protein expression on a post-transcriptional level.

Understanding the role of new isoforms and proteins with unknown functions in regulatory networks such as self-tolerance or tumorigenesis remains a challenge and awaits the improvement of assays for high-throughput screening of protein–protein interactions (Suzuki et al 2003) and functional states (i.e. phosphorylation) of proteins (Barrios-Rodiles et al 2005). For instance, EGF receptor (Araujo et al 2005) network modelling has the potential for improving non-small cell lung cancer therapy, but has not yet seen a breakthrough in the past 10 years. The Genomic Elements Viewer (*http://fantom32p.gsc.riken.jp/gev-f3/gbrowse/mm5*) and CAGE Analysis Viewer (*http://fantom31p.gsc.riken.jp/cage_analysis/mm5*) contain a potential tissue cell specific Egfr variant (AK087861) that lacks the first 20 exons and therefore the entire extracellular and transmembrane domains. The variant is produced by an alternative transcription start site supported by RIKEN 5' ESTs and CAGE tags. It remains to be seen whether the intracellular Egfr retains its autophosphorylation activity and how it affects the activation of the MAPK signalling pathway.

Role of non-canonical splicing products

Thus far we have considered only protein-coding variants generated by alternative splicing from one TU. The FANTOM3 annotation revealed 82 non-canonical in-frame and nine frame-shifted protein fusion candidates produced from two TUs. Probably, the fusions are the result of transcriptional readthrough of two closely positioned head-to-tail oriented TUs followed by *cis*-splicing of the hybrid pre-mRNA. Originally, the phenomenon was noted in a study of UEV1-KUA and predicted to be more widespread than expected (Thompson et al 2000).

In FANTOM3 most of the in-frame fusions were observed among paralogues, including members of the Ly49 natural killer cell receptor family members. Since the transcript fusions have the potential to alter signal transduction properties and/or MHC class I specificities they may represent yet another feature of the stochastic tissue cell-specific natural killer cell receptor activation that generate qualitative differences in immune responses. Since the fusion transcripts use the promoter of the first TU it is unclear when transcription results in two distinct transcripts or one fusion transcript. One possibility is that both pre-mRNAs are produced, but the expression level of splicing factors in a tissue cell determines whether the hybrid pre-mRNA is correctly spliced or subject to NDM.

Translational variation and gene expression regulation

Despite the importance of alternative splicing for the diversity of the proteome and regulation of gene expression two aspects of protein-coding mRNAs—alternative translation initiation and short peptides—have not yet been touched upon. Alternative translation initiation can generate from one transcript two or more protein isoforms. In the case of leaky scanning the ribosome bypasses the first ATG initiation codon and starts protein synthesis at a downstream ATG (Kozak 2002). Other non-canonical translation initiation modes include the use of a CTG start codon (Touriol et al 2003) or CAP-independent translation initiation by internal ribosomal entry sites (IRES) (Sachs 2000). Leaky scanning and reinitiation can result in the translation of an upstream open-reading frame (ORF) and a longer downstream ORF or two overlapping ORFs. In the latter case the alternative translation product can be a shorter in-frame variant or frame-shifted variant with different domain/motif, structure, membrane organization and functional properties.

The list of mammalian alternative translation-mediated isoforms, most of which were fortuitously discovered, is still small and awareness of the mechanism appears to be low. SWISS-PROT/TrEMBL contains 37 mouse and 49 human isoforms which are also incorporated along other literature-derived cases into the specialized Alternative Translation Initiation Database (ATID) (Cai et al 2005). XLalphas/Galphas and ALEX are two interacting proteins which are translated from overlapping frame-shifted ORFs of one transcript. Since the interaction is thought to affect the signal transduction of XLalphas (Klemke et al 2001) alternative translation initiation may have a broader biological significance as regulatory mechanisms in pathways. Likewise, it remains to be established whether alternative translation initiation is inducible, and when and why isoforms are produced by alternative translation, alternative splicing or both mechanisms. In addition, alternative translation products might be recognized as non-self and contribute to autoimmune reactions. Non-canonically translated tumour antigens (Wang et al 1996) that are recognized by CTL suggest such possibility.

Apart from IRES-mediated CAP-independent translation initiation which has been thoroughly studied in context of impaired CAP-dependent translation (i.e. G2/M phase of cell cycle, apoptosis, hypoxia) (Sachs et al 2000 and *http:// www.iresite.org*), alternative translation by CAP-dependent re-initiation or leaky scanning has not been systematically explored in mammals. Part of the reason is that ORF prediction software is optimized to identify canonical ORFs correctly, rather than non-canonical ORFs. In an attempt to identify potential overlapping ORFs we screened 66 799 FANTOM2 and non-RIKEN mouse cDNAs (GenBank141) with complete coding sequence information. Predicted alternative ORF candidates (2442) that shared at least 400 nucleotides with the known coding sequence were computationally categorized and annotated according to the type of ORF-overlap, Kozak context, ATG distance, cross-species conservation, protein motifs and solubility or transmembrane potential. The rule-based procedure yielded 1597 alternative ORFs of which 288 were assigned a high probability of being translated. Among these, candidates with homology to human proteins (12) and changes in protein membrane organization or domain structure (38) were of particular interest. One candidate, *BC003267*, encodes in its annotated main ORF a putative

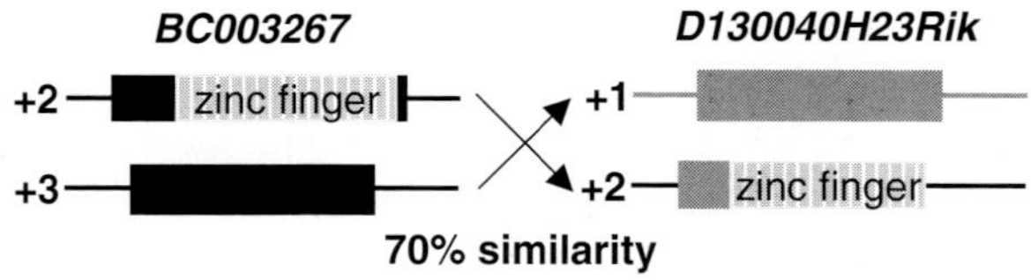

FIG. 1. Schematic representation of BC003267 and D130040H23Rik ORFs. The main and alternative ORF are shown as black and grey shaded boxes. Zinc-finger domains are shown as striped boxes. Arrows indicate sequence similarity relationships among main and alternative ORFs.

C2H2-type zinc-finger protein (Fig. 1). The downstream, +1-shifted alternative frame overlaps 80% with the main ORF and encodes a tissue-specific hypothetical protein that was confirmed with polyclonal antibodies (Kurochkin & Schönbach, unpublished work 2005).

Our preliminary data and reports of a growing number of small human proteins translated from short upstream ORFs (Oyama et al 2004) point towards an underestimation of alternative translation initiation. For instance, short upstream ORFs in CHOP and c-lck can suppress the translation of the downstream ORF, but not the expression of oncogenic forms (Jousse et al 2001). The alternative ORF sequence of *BC0003267* is similar to the main ORF sequence of *D130040H23Rik* on the same chromosome. Interestingly, the alternative ORF of *D130040H23Rik* may produce a zinc-finger protein that is similar in sequence to the main ORF of *BC0003267* (Fig. 1). The reciprocal relationships of the main and alternative ORFs could be part of a tissue-specific failsafe mechanism or repressor/activator switch.

Short peptides: yet another regulatory layer?

About 6% (2218) of the complete isoform protein set (*ftp://fantom.gsc.riken.jp/RTPS/fantom3_mouse/predicted_rtps/IP*) entries are shorter than 100 aa. However the majority of these peptides (50–100 aa), including antimicrobial peptides and neurohormones, are considerable longer (Frith et al 2006) than known endogenously encoded short peptides of 7–30 aa that are often tumour T-cell epitopes. These peptides appear to be produced by leaky-scanning (TRP1) (Wang et al 1996), IRES or perhaps from tissue cell-specific alternative splice variants (M-CSF) (Probst-Kepper et al 2001). Since the peptides were identified in tumour cell lines and not characterized beyond T-cell epitope level, their presence and function in normal cells is less clear.

In contrast, the discovery of humanin (24 aa) in an expression library screen for proteins that prevent aberrant amyloid peptide and presenilin triggering neuronal cell death in familial Alzheimer's disease, implicates a broader physiological role of short peptides (Hashimoto et al 2001). Indeed, humanin appears to be multifunctional and involved in different anti-apoptotic pathways by competing with β-amyloid in binding to the G protein-coupled formylpeptide receptor-like receptor (Ying et al 2004) or by binding to the proapoptotic Bax and preventing its translocation to the mitochondria (Guo et al 2003).

Another short peptide, cPR-3, is produced by the antisense strand of PRTN3 which encodes the autoantigen recognized by antineutrophil cytoplasmic autoantibodies in small-vessel vasculitis (Pendergraft et al 2004). The potential linkage of idiotypic dysregulation with antisense peptides should encourage the experimental and computational re-evaluation of the coding potential of transcripts currently designated as non-protein-coding. Mass spectrometric methods as applied in the Peptidome Project (*http://www.peptidome.org*) are still limited by the level of signal-noise ratio detection and ability to differentiate between proteolytic cleavage products and expressed peptides. Therefore, data mining methods that can overcome the weak signal of computationally predicted very short ORFs may have a lot to offer to support the identification of short peptides and uncovering their role as information mediators and regulators in the proteome.

Conclusions and future prospects

The transcriptome is a key mediator between the information encoded in the genome and cellular behaviour. Analyses of full-length cDNAs in combination with CAGE provide an essential base for decoding the underlying transcription regulatory networks. However, the information on its own is not sufficient to elucidate the variety of cause–effect relations in gene expression and cellular functions. Comprehensive insight into immune response regulation will depend not only on linking information beyond the transcriptome, but parallel sampling of epigenetic, protein–protein interaction, structure, signalling, metabolic and phenotypic data of specific cells at various conditions and developmental stages. Given the experimental and computational challenges to acquire data on protein or metabolic levels and analyse them in context of gene expression at reasonable speed and cost, it will be necessary to focus initially on a few model tissue cells. The Beta Cell Consortium (*http://www.betacell.org*), the Human Mast Cell Transcriptome Project (Saito et al 2001) and FANTOM3 data on macrophages already provide a rich resource on immune system relevant tissue cells, which can be built upon and expanded. Provided we can achieve a standardization of data at different levels of acquisition as a prerequisite for efficient data mining and visualization, immune regulatory network profiling and modelling may become applicable to medical diagnosis, prevention and treatment of immune system diseases.

References

Amarante MK, De Lucca FL, de Oliveira CE et al 2005 Expression of noncoding mRNA in human blood cells activated with synthetic peptide of HIV. Blood Cells Mol Dis 35:286–290

Araujo RP, Petricoin EF, Liotta LA 2005 A mathematical model of combination therapy using the EGFR signaling network. Biosystems 80:57–69

Bajic V, Tan SL, Christoffels A et al 2006 Mice and man: their promoter properties. PLoS Genet 2:e54

Barrios-Rodiles M, Brown KR, Ozdamar B et al 2005 High-throughput mapping of a dynamic signaling network in mammalian cells. Science 307:1621–1625

Brahmachary M, Schönbach C, Yang L et al 2006 Computational promoter analysis of mouse, rat and human antimicrobial peptide-coding genes. BMC Bioinformatics 7:58

Cai J, Zhang J, Huang Y et al 2005 ATID: a web-oriented database for collection of publicly available alternative translational initiation events. Bioinformatics 21:4312–4314

Carninci P, Kasukawa T, Katayama S et al 2005 The transcriptional landscape of the mammalian genome. Science 309:1559–1563

Chtanova T, Newton R, Liu SM et al 2005 Identification of T cell-restricted genes, and signatures for different T cell responses, using a comprehensive collection of microarray datasets. J Immunol 175:7837–7847

Frith MC, Forrest AR, Nourbakhsh E et al 2006 The abundance of short proteins in the mammalian proteome. PLoS Genet 2:e52

Guo B, Zhai D, Cabezas E et al 2003 Humanin peptide suppresses apoptosis by interfering with Bax activation. Nature 423:456–461

Hashimoto Y, Niikura T, Tajima H et al 2001 A rescue factor abolishing neuronal cell death by a wide spectrum of familial Alzheimer's disease genes and Abeta. Proc Natl Acad Sci USA 98:6336–6341

Hayashizaki Y 2003 Mouse genome encyclopedia project. Cold Spring Harb Symp Quant Biol 68:195–204

Jones RB, Wang F, Luo Y et al 2001 The nonsense-mediated decay pathway and mutually exclusive expression of alternatively spliced FGFR2IIIb and -IIIc mRNAs. J Biol Chem. 276:4158–4167

Jousse C, Bruhat A, Carraro V et al 2001 Inhibition of CHOP translation by a peptide encoded by an open reading frame localized in the chop 5' UTR. Nucleic Acids Res 29:4341–4351

Katayama S, Tomaru Y, Kasukawa T et al 2005 Antisense transcription in the mammalian transcriptome. Science 309: 1564–1566

Klemke M, Kehlenbach RH, Huttner WB 2001 Two overlapping reading frames in a single exon encode interacting proteins–a novel way of gene usage. EMBO J 20:3849–3860

Kozak M 2002 Pushing the limits of the scanning mechanism for initiation of translation. Gene 299:1–34

Krek A, Grun D, Poy MN et al 2005 Combinatorial microRNA target predictions. Nat Genet 37:495–500

Lewis BP, Green RE, Brenner SE 2003 Evidence for the widespread coupling of alternative splicing and nonsense-mediated mRNA decay in humans. Proc Natl Acad Sci USA 100:189–192

Ng P, Wei CL, Sung WK et al 2005 Gene identification signature (GIS) analysis for transcriptome characterization and genome annotation. Nat Methods 2:105–111

Oyama M, Itagaki C, Hata H et al 2004 Analysis of small human proteins reveals the translation of upstream open reading frames of mRNAs. Genome Res 14:2048–2052

Pendergraft WF 3rd, Preston GA, Shah RR et al 2004 Autoimmunity is triggered by cPR-3(105–201), a protein complementary to human autoantigen proteinase-3. Nat Med 10:72–79

Poy MN, Eliasson L, Krutzfeldt J et al 2004 A pancreatic islet-specific microRNA regulates insulin secretion. Nature 432:226–230
Probst-Kepper M, Stroobant V, Kridel R et al 2001 An alternative open reading frame of the human macrophage colony-stimulating factor gene is independently translated and codes for an antigenic peptide of 14 amino acids recognized by tumor-infiltrating CD8 T lymphocytes. J Exp Med 193:1189–1198
Purohit S, Kumar PG, Laloraya M et al 2005 Mapping DNA-binding domains of the autoimmune regulator protein. Biochem Biophys Res Commun 327:939–944
Ramkissoon SH, Mainwaring LA, Ogasawara Y et al 2006 Hematopoietic-specific microRNA expression in human cells. Leuk Res 30:643–647
Ravasi T, Suzuki H, Pang KC et al 2006 Experimental validation of the regulated expression of large numbers of non-coding RNAs from the mouse genome. Genome Res 16:11–19
Sachs AB 2000 Cell cycle-dependent translation initiation: IRES elements prevail. Cell 101:243–245
Saito H, Nakajima T, Matsumoto K 2001 Human mast cell transcriptome project. Int Arch Allergy Immunol 125:1–8
Shiraki T, Kondo S, Katayama S et al 2003 Cap analysis gene expression for high-throughput analysis of transcriptional starting point and identification of promoter usage. Proc Natl Acad Sci USA 100:15776–15781
Suzuki H, Saito R, Kanamori M et al 2003 The mammalian protein-protein interaction database and its viewing system that is linked to the main FANTOM2 viewer. Genome Res 13:1534–1541
Thomson TM, Lozano JJ, Loukili N et al 2000 Fusion of the human gene for the polyubiquitination coeffector UEV1 with Kua, a newly identified gene. Genome Res 10:1743–1756
Touriol C, Bornes S, Bonnal S et al 2003 Generation of protein isoform diversity by alternative initiation of translation at non-AUG codons. Biol Cell 95:169–178
Wang RF, Parkhurst MR, Kawakami Y et al 1996 Utilization of an alternative open reading frame of a normal gene in generating a novel human cancer antigen. J Exp Med 183:1131–1140
Ying G, Iribarren P, Zhou Y et al 2004 Humanin, a newly identified neuroprotective factor, uses the G protein-coupled formylpeptide receptor-like-1 as a functional receptor. J Immunol 172:7078–7085
Zavolan M, Schönbach C 2006 The contribution of alternative transcription and alternative splicing to the complexity of mammalian transcriptomes. In: Zhang W, Shmulevich I (eds) Computational and statistical approaches to genomics, 2nd edn, Springer Science + Business Media Inc. New York, p 351–380

DISCUSSION

Gros: As far as the percentage of transcripts that show alternative splicing goes, I had a couple of questions that relate to the primary material used for the analysis. For example, these alternatively spliced exons could also be viewed as incompletely spliced introns. The fact that on average you find them to be longer will go in that direction, at least in part. In terms of quality control for the RNA that is used as the initial material to create the cDNAs from which these sequence data are extracted, how much quality control is there to avoid contamination by partially spliced RNA? Are people looking at cytoplasmic RNA fractions versus total RNA fraction, for example? Also, I assume that when the sequences are analysed, a majority of transcripts look like the constitutively spliced exons, and then a small number may actually show a distinct pattern. In your computational analysis is there some sort of correction to avoid sampling bias? You might not be giving the same weight to things that show up occasionally, versus a majority of constitutively spliced exons. How would this affect your 65% of transcripts that are alternatively spliced?

Schönbach: Let's start with your question about the source of transcripts. The vast majority of transcript is cytoplasmic. In terms of sequencing errors, the quality is very high. A minority of transcripts is derived from nuclear RNA, and these transcripts are unprocessed.

Hume: Not only are these transcripts mainly cytoplasmic, but also these are full length cDNAs enriched with the cap trapping technology. Splicing is usually cotranscriptional. These are fully capped transcripts. Most heterogeneous nuclear RNA is actually not capped, because the cap has been processed off. The paper we have coming out in *Genome Biology* (Wells et al 2006) shows that all of the splice variants that were predicted by the RIKEN full-length transcripts were in fact confirmed as existing and regulated.

Gros: Do you think that all 65% are correct, and there is no sampling bias when you do the analysis?

Schönbach: The frequency of observed splice variants depends on the sequencing strategy and sampling. In fact, there is a sampling bias towards non-variant transcripts which was mentioned already in the FANTOM2 splice variation analysis (Zavolan et al 2003). Variation of internal exons is underestimated whereas variation of terminal exons is overestimated unless the bias is statistically corrected. Therefore the 65% is a conservative estimate. Interestingly, patterns associated with constitutive exons were weaker compared to variant exon-associated patterns. Overall, the effect of sampling bias on identifying patterns predictive of variant exons appears to be small.

Hume: Actually, the sampling bias is in the other direction. The whole program was designed to sequence one representative transcript of each transcriptional unit. The first triage is to sequence the 5′ and 3′ end first, and say we've seen that before and so don't bother sequencing the rest. If anything, the internal exon alternate splicing has been substantially undersampled at this time point because it was deliberately avoided. The first priority of this project was to get a representative transcript for every transcriptional unit. The project isn't designed to sequence alternative splicing; it is designed to avoid it.

Alexander: What patterns can you see emerging about the dynamic regulation of these alternate splicing forms? The fascinating concept is that these will be dynamically varying in cells, and this will allow cells to express different proteins and respond differently to different stimuli. Is it clear that there is a large amount of dynamic regulation of differently spliced forms? From the patterns that you are seeing, is there any coordinated way in which this appears to be happening? For example, in a data set like that David Hume has, with macrophages responding to LPS, can you see alternate splicing patterns that change with time? It is difficult to understand how this might be done on a gene-by-gene rather than global basis.

Schönbach: In the present analyses the dynamic aspect of splicing regulation has not been covered.

Hume: We have looked at this. In our forthcoming *Genome Biology* paper we have splicing-specific arrays (Wells et al 2006). It is a set of array tags down the transcript that detect the different forms differentially. And they are differently regulated across a time of activation.

Alexander: What sort of mechanism are you thinking of here? We are used to thinking of splicing as being a splicing machine recognizing the specific acceptor or donor sites.

Schönbach: Acceptor and donor site signals are necessary but not sufficient for the definition of exon/intron boundaries and splicing. Additional, often stronger patterns in introns and exons are recognized by proteins of the splicing complex. For example splicing regulatory proteins recognize *cis*-elements that enhance or silence splicing. Assuming that the splicing complex proteins can recognize combinations of potential splice sites and associated patterns such as splice enhancers/silencers, the outcome of alternative splicing is probably determined by the concentration of proteins and their rate of signal or pattern recognition.

Hume: Some of the components of the splicing apparatus are themselves transcriptionally regulated in the profile of macrophage activation. This would be one way it could happen: just having more of the critical protein available to allow an exon to be included with greater probability. This is largely virgin territory. One of the discoveries of the RIKEN project is that a large number of exons themselves have promoter activity. This is a gene-specific phenomenon. If you look at the collagen gene, *COL3A1*, this has no promoter activity over any of the exons. In albumin, on the other hand, which is comparably expressed, all of the exons have quite strong transcriptional promoter activity. To speculate whether the transcription apparatus has a role in controlling levels of expression and splicing is a new player in the whole process.

Schönbach: For the 5′ terminal exon, the transcription initiation influences alternative splicing. This is probably not the case for skipping of middle exons.

Vinuesa: What has been done with the analysis of the non-coding RNAs? Do they parallel in distribution the density of transcription units? Or do you see biases and enrichment according to family functions?

Schönbach: What I showed in the presentation was an analysis of sense/anti-sense transcripts which includes only a subset of non-coding RNAs. The analysis of all non-coding RNAs in the FANTOM3 paper (Carninci et al 2005) didn't report any bias with regard to chromosomal distribution or GC content. However about half of the non-coding RNA overlapped with protein-coding RNAs. Among these non-coding RNAs many were shown to overlap with 3' UTRs of protein-coding RNAs, implying some regulatory role. The remaining non-coding RNAs were single transcript clusters or singletons

References

Wells CA, Chalk AM, Forrest A 2006 Alternate transcription of the Toll-like receptor signaling cascade. Genome Biol 7:R10

Zavolan M, Kondo S, Schonbach C et al 2003 Impact of alternative initiation, splicing, and termination on the diversity of the mRNA transcripts encoded by the mouse transcriptome. Genome Res 13:1290–1300

Molecular pathways for lymphangiogenesis and their role in human disease

Steven A. Stacker, Rae H. Farnsworth, Tara Karnezis, Ramin Shayan, Darrin P. Smith, Karri Paavonen, Natalia Davydova, Carol Caesar, Rachael Inder, Megan E. Baldwin, Bradley K. McColl, Sally Roufail, Richard A. Williams*, Richard A. Hughes†, Kari Alitalo‡ and Marc G. Achen

*Angiogenesis Laboratory, Ludwig Institute for Cancer Research, PO Box 2008, Royal Melbourne Hospital, 3050, Victoria, Australia, *Department of Pathology, Faculty of Medicine, Dentistry and Health Sciences and †Department of Pharmacology, University of Melbourne, Victoria, Australia and ‡Molecular/Cancer Biology Laboratory and Ludwig Institute for Cancer Research, Biomedicum Helsinki and Helsinki University Central Hospital, University of Helsinki, Helsinki, Finland*

Abstract. The lymphatic network functions to return fluid, cells and macromolecules to the circulation. Recent characterization of growth factors that control the growth and development of the lymphatics, and markers which specify lymphatic endothelial cells have enhanced our understanding of this system. Members of the VEGF family of factors are key regulators of these vessels with VEGF-C/VEGF-D and VEGFR-3 being the best validated signalling pathways in lymphangiogenesis. The study of these molecules in various pathologies has shown that they are important in the processes of cancer metastasis and in the formation of lymphoedema. Knowledge of these molecular pathways allows for the generation of modulators of these pathways which could form the basis of novel therapeutic approaches.

2007 Decoding the genomic control of immune reactions. Wiley, Chichester (Novartis Foundation Symposium 281) p 38–49

The lymphatic system is a series of vessels for transporting back to the circulation, the fluid, cells and macromolecules that escape from the blood vascular system. It is therefore an essential part of the circulatory system. In normal physiology the lymphatics therefore play a key role in fluid homeostasis and immune cell trafficking. In pathological conditions aberrant lymphatic vessel function is associated with conditions such as lymphoedema and cancer (Baldwin et al 2002).

In the past decade the discovery of growth factors and receptors which can stimulate the proliferation and differentiation of lymphatic endothelial cells (lymphangiogenesis) has facilitated renewed investigation of the lymphatic network and its role in disease (Oliver & Detmar 2002). The best validated signalling systems that control lymphangiogenesis involve members of the vascular endothelial growth factor (VEGF) family of soluble, secreted proteins and their associated cell surface receptors (VEGFRs). The family consists of VEGF or VEGF-A, VEGF-B, VEGF-C, VEGF-D and placenta growth factor (PlGF). There are also VEGF members produced by viruses and in the venom of snakes. These growth factors signal through interaction with cell surface receptor tyrosine kinases (RTK) or growth factor receptors. Three RTKs are involved in binding and signalling for the VEGFs; VEGFR-1, VEGFR-2 and VEGFR-3. These receptors are expressed predominately on endothelial cells and their stimulation drives the processes of angiogenesis and lymphangiogenesis. VEGFR-2 acts as the primary mitogenic receptor for blood vascular endothelial cells (i.e. promoting angiogenesis) and VEGFR-3 as the prime mitogenic receptor for lymphatic endothelial cells (i.e. promoting lymphangiogenesis). These growth factors also utilize members of the Neuropilin family of cell surface receptors, and this provides an interesting diversity for generating responses such as vascular permeability. It also provides a link to the neural system in which VEGFs also play a role and may provide a critical biological link between two vital systems in animals; the vascular system and the nervous system.

A subfamily of the VEGFs, consisting of VEGF-C and VEGF-D, appears to be the most potent lymphangiogenic growth factors. These growth factors have high homology to the angiogenic factor VEGF through a conserved cysteine-knot motif, and also have N- and C-terminal propeptide extensions which are cleaved by proteases to generate mature forms capable of high affinity binding to both VEGFR-2 and VEGFR-3 (Joukov et al 1996, 1997, Achen et al 1998, Stacker et al 1999). The unprocessed forms of VEGF-C and VEGF-D have significantly reduced capacity to bind the receptors (40- and 300-fold reduced affinity for VEGFR-3 and VEGFR-2, respectively, in the case of VEGF-D). Therefore activation by proteases is essential for regulating the binding of the growth factors to the two receptors and the absence of proteolytic processing may prevent VEGF-C and VEGF-D from inducing angiogenesis. Currently two proteases have been characterized as being capable of processing VEGF-C and VEGF-D *in vitro*, namely plasmin (McColl et al 2003) and the proprotein convertase Furin (Siegfried et al 2003). Further studies are progressing to determine the key enzymes for *in vivo* cleavage of these growth factors and to determine which are required for regulating the activity in developmental versus pathological contexts.

The recent discovery and characterization of the lymphangiogenic growth factors has provided an understanding of the molecular mechanisms controlling the growth of lymphatic vessels. In addition, other markers have been identified with specificity for lymphatic endothelium and this has facilitated much recent research including characterizing lymphatic vessels in developmental and pathological contexts and isolating and characterizing lymphatic endothelial cells. The key markers have been podoplanin, a molecule found on kidney podocytes that is also localised on lymphatic endothelial cells. Its resistance to protease treatment makes it a useful tool for purifying populations of lymphatic endothelial cells. The hyalauronin receptor LYVE-1 (Banerji et al 1999, Jackson et al 2001) and VEGFR-3 have also been used to identify lymphatic vessels by immunohistochemistry.

The key role of the VEGF-C/VEGF-D/VEGFR-3 pathway in lymphangiogenesis and lymphatic development has been well demonstrated initially using *in vitro* techniques and more recently using a variety of animal models (Jeltsch et al 1997, Achen et al 1998, Dumont et al 1998, Makinen et al 2001a, Veikkola et al 2001, Stacker et al 2002, Karkkainen et al 2004). Initial studies showed that some of these proteins were capable of driving the proliferation of cultured endothelial cells, and inducing lymphangiogenesis in the chick chorioallantoic membrane (CAM) assay. Over-expression of VEGF-C or VEGF-D in the skin of transgenic animals showed their capacity to drive lymphangiogenesis *in vivo*. More recent studies in which VEGF-C and VEGF-D have been introduced into animal tissues using adenoviral vectors have shown their ability to drive both angiogenesis and lymphangiogenesis, and have highlighted their potential utility for the generation of new blood vessels and lymphatic vessels in humans in clinical settings.

The role of lymphangiogenic growth factors in cancer is an important discovery which has come out of the characterization of these important developmental proteins (Achen et al 2005). A series of initial studies from a number of groups showed that over-expression of the genes encoding the lymphangiogenic factors VEGF-C and VEGF-D in human tumour xenografts was capable of driving tumour lymphangiogenesis (Karpanen et al 2001, Mandriota et al 2001, Skobe et al 2001, Stacker et al 2001). Significantly this was also associated with an increased capacity of these tumours to spread to regional lymph nodes, indicating that the expression of the growth factors was somehow contributing to tumour cells entering the lymphatic network and metastasizing to a distant site. In our own studies we had shown that the spread could be inhibited by the application of an inhibitory monoclonal antibody to the growth factor (VEGF-D), demonstrating a proof-of-principle for anti-metastatic therapy via blocking the interaction of VEGF-D and its receptors. Analysis of these growth factors in human tumours was an important validation of the hypothesis that they could be important for driving at least one component of metastasis in human tumours; that of spread to regional lymph nodes. There are now over 50 studies which have used either RT-PCR or immunohistochemistry to quantitate the levels of VEGF-C and VEGF-D in the primary lesions of patients with epithelial tumours and correlate these with patient history (Stacker et al 2004b). Solid tumours expressing VEGF-C or VEGF-

D tend to exhibit spread to regional lymph nodes as a relatively early event in their progression. In general the results of these studies have been supportive of the conclusion that the expression of VEGF-C and VEGF-D in human tumours correlates with lymphogenous spread of the tumour and poor patient outcome. In some cases the expression of the growth factor has provided independent prognostic information on the outcome of the patients (White et al 2002, Yokoyama et al 2003). These studies have validated the initial animal models and suggest that the growth factors play a key role in assisting the spread of tumour cells from the primary tumour to the lymph nodes. What remains unknown at this stage is the exact mechanism by which these growth factors act in cancer. Initially it was thought that increased lymphangiogenesis could promote better access to the pre-existing lymphatic network, although it may be that this could be an unrelated activity and that the growth factors work to facilitate adhesion/entry to the pre-existing network or simply act to enlarge the calibre of lymphatic vessels allowing greater flow to regional nodes (Padera et al 2002, He et al 2005).

Dysfunction of the lymphatics can cause the accumulation of fluid in tissues, a condition called lymphoedema. In humans this condition can arise as a consequence of infection, surgery, radiation treatment or physical damage to a limb where the lymphatic network is removed or disrupted. In addition patients can suffer primary lymphoedema in which there is a genetic mutation affecting the development or function of the lymphatic system (Baldwin et al 2002). Lymphoedema is a progressive disease and left untreated can develop from the accumulation of small volumes of fluid in the periphery to an advanced condition of permanent fluid accumulation, tissue damage, infection and immobility of the affected limb. In some patients with primary lymphoedema, Milroy's disease, the mutation affecting these individuals has been mapped to the receptor for the lymphangiogenic receptor VEGFR-3 (Karkkainen et al 2000). This is supported by experimental evidence in mouse models where the VEGFR-3 signal transduction pathway has been altered (Makinen et al 2001b).

Both cancer and lymphoedema are potential targets for treatment via the targeting of the VEGF-C/VEGF-D/VEGFR-3 pathway (Stacker et al 2004a). In cancer, blocking the activity of the growth factors and receptor is a viable way to target this pathway. That Avastin, a humanized monoclonal antibody to VEGF, is capable of blocking tumour angiogenesis and provides a survival advantage in patients with advanced colorectal cancer is demonstration that this type of approach is possible (Kabbinavar et al 2005). Furthermore, the group of small molecule tyrosine kinase inhibitors of VEGFR-2 (Achen et al 2005) may also have activity against the structurally-related receptor VEGFR-3 and therefore exhibit anti-lymphangiogenic activity. Such small molecules and monoclonal antibodies to VEGF-C and VEGF-D are currently being pursued for their capacity to inhibit these signalling pathways in humans.

Lymphoedema is a disease which currently has no molecular-based treatment. Theoretically one could treat this condition by expressing lymphangiogenic growth factors in the affected limb to drive the formation of new lymphatic channels or repair of damaged lymphatic vessels. This has been performed in animal models, with new vessels being able to alleviate the lymphoedema. Currently, the complexity of applying these factors in a localized fashion is being studied, with future clinical trials a possibility.

Even though the VEGF family play a key role in regulating the development of the lymphatic network, it is clear that other molecules are capable of promoting lymphangiogenesis, either directly or indirectly (Hirakawa et al 2005, Kajiya et al 2005). As our knowledge of the function of lymphatic vessels broadens it is likely that we will see, akin to blood vessels that the lymphatic vessels serve not just as passive pipes to clear fluid but as dynamic interactive surfaces which help to regulate the functions of cells which flow through them. Our growing understanding of the molecules that control lymphangiogenesis allows us to design more specific drugs with which to manipulate the relevant signalling pathways (Achen et al 2005, Stacker et al 2004a). Modulating these pathways and other molecules with specificity to the lymphatic system could offer alternative treatments for a number of important clinical conditions, such as cancer and lymphoedema.

References

Achen MG, Jeltsch M, Kukk E et al 1998 Vascular endothelial growth factor D (VEGF-D) is a ligand for the tyrosine kinases VEGF receptor 2 (Flk1) and VEGF receptor 3 (Flt4). Proc Natl Acad Sci USA 95:548–553

Achen MG, McColl BK, Stacker SA 2005 Focus on lymphangiogenesis in tumor metastasis. Cancer Cell 7:121–127

Baldwin ME, Stacker SA, Achen MG 2002 Molecular control of lymphangiogenesis. Bioessays 24:1030–1040

Banerji S, Ni J, Wang S-X et al 1999 LYVE-1, a new homologue of the CD44 glycoprotein, is a lymph-specific receptor for hyaluronan. J Cell Biol 144:789–801

Dumont DJ, Jussila L, Taipale J et al 1998 Cardiovascular failure in mouse embryos deficient in VEGF receptor-3. Science 282:946–949

He Y, Rajantie I, Pajusola K et al 2005 Vascular endothelial cell growth factor receptor 3-mediated activation of lymphatic endothelium is crucial for tumor cell entry and spread via lymphatic vessels. Cancer Res 65:4739–4746

Hirakawa S, Kodama S, Kunstfeld R, Kajiya K, Brown LF, Detmar M 2005 VEGF-A induces tumor and sentinel lymph node lymphangiogenesis and promotes lymphatic metastasis. J Exp Med 201:1089–1099

Jackson DG, Prevo R, Clasper S, Banerji S 2001 LYVE-1, the lymphatic system and tumor lymphangiogenesis. Trends Immunol 22:317–321

Jeltsch M, Kaipainen A, Joukov V et al 1997 Hyperplasia of lymphatic vessels in VEGF-C transgenic mice. Science 276:1423–1425

Joukov V, Pajusola K, Kaipainen A et al 1996 A novel vascular endothelial growth factor, VEGF-C, is a ligand for the Flt4 (VEGFR-3) and KDR (VEGFR-2) receptor tyrosine kinases. EMBO J 15:290–298

Joukov V, Sorsa T, Kumar V et al 1997 Proteolytic processing regulates receptor specificity and activity of VEGF-C. EMBO J 16:3898–3911

Kabbinavar FF, Hambleton J, Mass RD, Hurwitz HI, Bergsland E, Sarkar S 2005 Combined analysis of efficacy: the addition of bevacizumab to fluorouracil/leucovorin improves survival for patients with metastatic colorectal cancer. J Clin Oncol 23:3706–3712

Kajiya K, Hirakawa S, Ma B, Drinnenberg I, Detmar M 2005 Hepatocyte growth factor promotes lymphatic vessel formation and function. EMBO J 24:2885–2895

Karkkainen MJ, Ferrell RE, Lawrence EC et al 2000 Missense mutations interfere with VEGFR-3 signalling in primary lymphoedema. Nat Genet 25:153–159

Karkkainen MJ, Haiko P, Sainio K et al 2004 Vascular endothelial growth factor C is required for sprouting of the first lymphatic vessels from embryonic veins. Nat Immunol 5:74–80

Karpanen T, Egeblad M, Karkkainen MJ et al 2001 Vascular endothelial growth factor C promotes tumor lymphangiogenesis and intralymphatic tumor growth. Cancer Res 61:1786–1790

Makinen T, Veikkola T, Mustjoki S et al 2001a Isolated lymphatic endothelial cells transduce growth, survival and migratory signals via the VEGF-C/D receptor VEGFR-3. EMBO J 20:4762–4773

Makinen T, Jussila L, Veikkola, T et al 2001b Inhibition of lymphangiogenesis with resulting lymphedema in transgenic mice expressing soluble VEGF receptor-3. Nat Med 7:199–205

Mandriota SJ, Jussila L, Jeltsch M et al 2001 Vascular endothelial growth factor-C-mediated lymphangiogenesis promotes tumour metastasis. EMBO J 20:672–682

McColl BK, Baldwin ME, Roufail S et al 2003 Plasmin activates the lymphangiogenic growth factors VEGF-C and VEGF-D. J Exp Med 198:863–868

Oliver G, Detmar M 2002 The rediscovery of the lymphatic system: old and new insights into the development and biological function of the lymphatic vasculature. Genes Dev 16:773–783

Padera TP, Kadambi A, di Tomaso E et al 2002 Lymphatic metastasis in the absence of functional intratumor lymphatics. Science 296:1883–1886

Siegfried G, Basak A, Cromlish JA 2003 The secretory proprotein convertases furin, PC5, and PC7 activate VEGF-C to induce tumorigenesis. J Clin Invest 111:1723–1732

Skobe M, Hawighorst T, Jackson DG et al 2001 Induction of tumor lymphangiogenesis by VEGF-C promotes breast cancer metastasis. Nat Med 7:192–198

Stacker SA, Stenvers K, Caesar C et al 1999 Biosynthesis of vascular endothelial growth factor-D involves proteolytic processing which generates non-covalent homodimers. J Biol Chem 274:32127–32136

Stacker SA, Caesar C, Baldwin ME et al 2001 VEGF-D promotes the metastatic spread of tumor cells via the lymphatics. Nat Med 7:186–191

Stacker SA, Achen MG, Jussila L, Baldwin ME, Alitalo K 2002 Lymphangiogenesis and cancer metastasis. Nat Rev Cancer 2:573–583

Stacker SA, Hughes RA, Achen MG 2004a Molecular targeting of lymphatics for therapy. Curr Pharm Design 10:65–74

Stacker SA, Williams RA, Achen MG 2004b Lymphangiogenic growth factors as markers of tumor metastasis. APMIS 112:539–549

Veikkola T, Jussila L, Makinen T et al 2001 Signalling via vascular endothelial growth factor receptor-3 is sufficient for lymphangiogenesis in transgenic mice. EMBO J 20:1223–1231

White JD, Hewett PW, Kosuge D et al 2002 Vascular endothelial growth factor-D expression is an independent prognostic marker for survival in colorectal carcinoma. Cancer Res 62:1669–1675

Yokoyama Y, Charnock-Jones DS, Licence D et al 2003 Vascular endothelial growth factor-D is an independent prognostic factor in epithelial ovarian carcinoma. Br J Cancer 88:237–244

DISCUSSION

Mackay: From an immunologist's perspective, the main purpose of lymphangiogenesis is to create lymphatic vessels for immune cells to promote lymphatic vessels from a pathogenic site. When I was studying sheep 15–20 years ago we injected antigen into the skin and saw incredible formation of lymphatic vessels to a node and from a node. It makes me wonder whether it is the tumour that is crucial in promoting lymphangiogenesis or in fact the immune response to the tumour. What cell types actually produce VEGFs?

Stacker: That is a good question. Clearly, within the human, we suspect it is going to be other cell types producing VEGF. Not only the tumour cells, but also the supporting stroma and macrophages are possibly contributing to the level of factors that are there. This is something that some companies (e.g. Genentech) are working on. What other components of stroma are contributing to this, and how can those factors be eliminated? When we originally got the data it looked as though there was a correlation between lymphangiogenesis and the rate of tumour spread. But now it is becoming clear that it may be the activity of these factors on pre-existing vessels further downstream that might be the critical determinant of metastases. Therefore proliferation of vessels occurs, but it could be just a bit of a bystander. How those factors then act on the molecular expression of downstream vessels may be the most important thing.

Goodnow: What about the idea that macrophages are one of the main sources of the VEGFs?

Stacker: That's a possibility. We see macrophages expressing VEGF-C and D. There has been a lot of controversy in the field about whether bone marrow-derived cells or resident macrophages in tissues might be contributing to the secretion of these factors. There is a lot of talk about bone-marrow derived cells that may be pre-seeding some of these metastatic lesions. There was a recent paper in *Nature* in which these VEGFR-1 positive cells were coming out of the bone marrow and defining those organs that cells are going to metastasize to (Kaplan et al 2005). There has always been a lot of controversy about whether these are resident macrophages that have been stimulated to appear, or whether they are coming from the bone marrow. The macrophage may be the central orchestrator of the whole process.

Cyster: Some papers have suggested that macrophage lineage cells can contribute to the vessel itself (Maruyama et al 2005). Is this well founded?

Stacker: These are data suggested by one or two high quality papers. In melanoma there was talk that you could get dedifferentiation of a whole host of cell types and they would incorporate back into the vessel. Two or three years ago there was a lot of interest that tumour cells would dedifferentiate back into forming endothelial cells. There was some thought at the time that macrophages may be involved there. The data are fairly thin still. It is a bit like the cancer stem cell: it is an appealing idea, without a lot of evidence at this stage.

Cyster: When you are forming a new lymphatic vessel in an adult mouse or human, is it thought to be a budding event, or is it coalescence of some type of cell?

Stacker: Generically, people have thought of it as being a budding or sprouting event. But I think we know now that it is probably not that straightforward. There are other techniques of intussusception and circulating progenitors which could contribute to vessel formation in certain contexts. Sprouting will probably be the predominant feature, but in individual contexts there may be variations on this theme. It seems that in problems such as cardiovascular disease, circulating progenitors may be important at building up bulk in a plaque area. In other situations, say tumour angiogenesis where you think the same process might occur, you don't detect that. Hence, there may be very individual variation on how this occurs, and I think we still haven't understood this.

Britton: In circumstances where you get tertiary lymphoid organs, such as chronic antigenic stimulation in autoimmune disease or infection, does blocking the lymphangiogenic factors actually block the characteristic immune response?

Stacker: I don't think it has been done, even though these would be straightforward experiments.

Britton: Presumably you get expression of new lymphatics in those systems?

Stacker: I would imagine so. There have been studies showing that in various transplantation models the rejection previously thought to be supported by the blood vascular system is now occurring via the lymphatics. When people use specific markers for lymphatics, certain kidney rejection models show that the vessels are predominately lymphatic in their nature. In most of the other situations, I don't think people have examined those models in detail. This is how new the field is.

Britton: What do you mean when you say 'lymphatic in nature'? Do you mean the vessels are recruited there?

Stacker: The vessels that are recruited or at least expand in the rejection process have strong expression of markers for lymphatic vessels. It was previously thought there was an infiltration of angiogenic vessels, and this was part of the rejection process. It appears now as though a high proportion of those vessels have lymphatic markers and appear to be behaving as lymphatic vessels. This is only one of a few immune regulated model that has been reasonably studied, apart from the tumour models which have been well characterized.

Hume: This conversion question of the myeloid to endothelial lineage is well supported by the recent literature. If you go back a little way it is worth remembering that in someone with Philadelphia chromosome CML, all the endothelial cells in both the vasculature and the lymphatics have the Philadelphia chromosome. So they are essentially of bone marrow origin, and are presumably arising from the same stem cell as the myeloid tumour. You can take pure CD14-positive human peripheral blood monocytes and turn them into the most beautiful tubes. Perhaps we need to go back to the reticuloendothelial system. You draw your diagrams of vessels with an ill-defined stroma, but the actual architecture of the vessel is that the macrophages spiral down the outside of the endothelial cells. There is a roughly one-to-one ratio of macrophage to endothelial cell in this structure. An interesting question is whether the immune response happens and then the lymphatics follow is an interesting question. I would tend to think that the immune response causes the lymphatic sprouting rather then being a consequence of it.

Stacker: There are a lot of contradictory data out there on bone marrow-derived cells generating blood vessels. Studies by Kari Alitalo's group in Helsinki and Glen Begley's group at Amgen have suggested that labelled bone marrow cells do not contribute to the endothelial cells in tumours. It may be context dependent. Also it is possible that bone marrow derived cells may provide essential factors to promote vessel formation in tumours rather than be the origin of the endothelial cells themselves. These are questions which still need to be addressed.

Turner: Are these factors involved in the maintenance of lymphatics as well as their generation?

Stacker: That is what we suspect. The data in this area are not complete. Certainly with VEGF in clinical models, they originally thought it was just a proliferative factor. But in some of the Genentech studies where they inhibited VEGF with antibody, they found that it was playing a key survival role in the lung. In patients treated with anti-VEGF antibody, the endothelial cells in the lung died off and the patients developed emphysema. In some contexts I think the lymphangiogenic factors such as VEGF-C and VEGF-D will be required for the survival of lymphatic endothelial cells and the vessels.

Goodnow: What is the evolutionary picture? There are important differences in the lymph node arrangement in the chicken, aren't there?

Kaufman: All these animals have lymphatics, but it is only the mammals that have lymph nodes.

Hume: One of the things I noticed in the data set I put up earlier is that VEGF-D is expressed at high levels in bone osteoclasts. There are also publications showing that VEGF-A acts on osteoclasts. The CSF1-deficient mouse gets better with age because of the actions of VEGF-A. Do you envisage roles in Ca^{2+} metabolism for VEGFs?

Stacker: Frankly, no. We have done the VEGF-D knockout and we don't see any bone pathologies.

Hume: It might be interesting to stress the bone system and see what happens in those animals.

Goodnow: To what extent do T cells make VEGF-D or A?

Stacker: Lung and skin is where we predominantly see VEGF-D. VEGF-C is a little more broadly expressed on the haematopoietic lineages. There is a report showing a role in B cell function, I think, for VEGF-C.

Alexander: In the clinical context, for the use of these factors in lymphoedema or restenosis, you were talking about the need to develop vectors and expression systems. Is your view that there is a need for local expression of these factors in these contexts? Is this a systemic toxicity issue, or one of efficiency?

Stacker: It's because you probably need prolonged expression of the protein in a regional area to get an effect. In the animal models, if you inject protein *in vivo* it will probably have an effect for a day or two, but you most likely require prolonged expression in that region for it to work well. From the clinical trials that have been performed, the adenovirus works well in a local setting and doesn't seem to have any systemic effects. It may be that you get a good level of production in the regional area and the protein doesn't drift too far from here, so its effect three or four centimetres away doesn't seem to be major. This may be the ideal system for using adenovirus.

Alexander: Would you expect significant side effects if it was systemically administered?

Stacker: You never really know with these types of factors. You might drive abnormalities of blood and lymphatic vessels at peripheral sites. If you look at the link with cancer, you might think that injecting yourself systemically with VEGF-C or -D might not be the best thing to do if it is going to promote vessel growth which may assist the growth and spread of cancer. We do not see this as a problem when using a local application of the factor. The cancer data show that almost all healthy people at autopsy will have tumour lesions, it is just that those lesions don't get beyond an extremely small size. They haven't developed that angiogenic component to get them past that critical small size. When some of the other data come through from Genentech with the Avastin trials, it will be interesting to see if it has a longer term effect on patients with minimal disease.

Cyster: One general thing that puzzles me is that lymphatic density in different tissues can be very different. In the thymus, for example, there is minimal lymphatic presence. How does fluid return occur in tissues that have minimal lymphatics?

Stacker: That's a good question. The blood vessel system is very specific in different organs. We presume that lymphatics will be the same, but at this stage we know very little about them. We have just a handful of markers. We need to develop to a point where we have 50 markers so we can start to subtype the various vessels in different organs, much like the CD system for haematopoietic cells. This is something we are looking at now.

Cyster: Lymphatics evolved to return fluid back to the circulation, didn't they? In which case, immune cells have taken advantage of the system.

Stacker: Developmentally, there are many interesting theories. I recently attended a conference in Miami where someone proposed that the VEGFs are all neuronal factors, and they have now been co-opted by the immune system.

Cyster: Presumably the fluid return function is very basic and goes back much earlier than cellular trafficking.

Stacker: That is possible. We now know that the zebrafish has a functioning lymphatic system with both a fluid clearing ability and circulating immune cells. On this basis, perhaps the cells have come first and the fluid clearing has come about as creatures have become more vertical, with an increase in fluid pressure. In the mouse and human there is a big difference in the lymphatic system because the mouse is not as vertically challenged as the human.

Britton: What happens when you remove lymph nodes? Many of the patients with lymphoedema have had their lymph nodes removed, so what happens to the distal lymphatics?

Stacker: In some cancer patients surgeons do lymph node (e.g. axillary) clearance, taking out a set of nodes. My understanding is that the vessels stay there but in many cases the vessels do not reconnect well. This is often where the lymphoedema arises, with two sections of the lymphatic drainage system separated.

Alexander: In the sheep, if we take out a lymph node, the lymphatics disappear.

Hume: This is the standard mesenteric lymphadenectomy. You strip the whole mesenteric lymph node chain out and get afferent lymph into the thoracic duct.

Stacker: There are different patterns of lymphatic drainage areas in the animal, and sometimes those areas will be compensated from other regions if disturbed. Therefore if the fluid won't go through a certain area, you may get an abnormal connection. This can occur in some cases, but in many cases if you effectively remove the lymphatics it can't track its way back. There are plenty of cases from cancer patients where you eliminate the connection and it doesn't reconnect. This may have to do with particular calibres of lymphatic vessels which themselves can't react to the change.

Hume: It could also be because of certain surgical techniques.

Stacker: Historically cancer surgery has typically been crude and hasn't appreciated the lymphatic network. This is where many of these problems derive from. In the lymphoedema patient support groups I have talked with, there was a heightened awareness in these patients that the surgeons sometimes don't do the best things for them. There is also a good appreciation from these people who are suffering from lymphoedema about how to care for their condition.

Belz: Is there much known about the molecules that guide the budding of new lymphatics?

Stacker: There is some but it is still quite incomplete at this stage. There is a group of molecules that are associated with the nervous system, like the semaphorins which clearly play a role. If you go to the vascular system which has been much more studied, the information there is developing. Molecular systems like the Ephs and the ephrins seem to play a major role in determining vascular boundaries and subtypes, so chances are these sorts of families will be significant in the lymphatic system.

Cook: It is often said that the white pulp of the spleen lacks lymphatic drainage. Does that hold up when you use these other markers? And does it persist when you induce an immune response?

Stacker: I don't know the answer. The markers that define the lymphatics have only come on the scene in the last three years. People haven't performed these fundamental experiments in detail yet.

Cyster: There is minimal staining with LYVE-1 in the spleen or thymus while lymph nodes stain extensively. In human there is more data to support the presence of a lymphatic drainage from the human spleen.

Stacker: I don't know the full extent of the data.

Cyster: People normally think of the venous sinusoids being the return route from the spleen. It would be interesting to look at human spleen more extensively with the new markers.

Cook: It is at odds with the immune response inducing these vessels in other organs.

Cyster: Perhaps it is suppressed in the spleen. Certainly, if you induce a response in the lymph node you can increase lymphatic vessel development.

Goodnow: Is there any evidence that microbes have developed ways of blocking growth, especially some of the more specialized pathogens? This could be a way of doing pharmaceutically what was done in the sheep.

Stacker: It hasn't been shown but, like the angiogenic system, there will probably be a large numbers of inhibitors of the process. These will become just as important to understand as the promoters of vessel formation. I also imagine the same will hold true for the lymphatic network.

Goodnow: So nothing has cropped up in the pox virus or herpes virus genomes?

Stacker: Not that I've seen. Although we do know that VEGF-like molecules which promote vessel formation are made by some viruses.

Cyster: They express molecules that inhibit chemokines, but whether the lymphangiogenesis stops I don't think has been assessed.

References

Kaplan RN, Riba RD, Zacharoulis S et al 2005 VEGFR1-positive haematopoietic bone marrow progenitors initiate the pre-metastatic niche. Nature 438:820–827

Maruyama K, Ii M, Cursiefen C et al 2005 Inflammation-induced lymphangiogenesis in the cornea arises from CD11b-positive macrophages. J Clin Invest 115:2363–2372

GENERAL DISCUSSION I

Goodnow: We have some time for general discussion of any unifying themes or particular issues that people want to discuss.

Maizels: One thing that has impressed me in our discussions is the level of tissue specificity. How far along are we in terms of identifying tissue-specific transcription factors? Are these responsible for the alternative start forms that David Hume talked about? Can this be pieced together or is this still a black box?

Hume: One of the problems with most array platforms is that the sensitivity is not sufficient to get all the transcription factors that are expressed. Affymetrix is quite unreliable just around the critical window of transcription factors that we know are important. This is why the quantitative real-time approach has been interesting. In macrophages we discovered that almost 70% of the total transcription factors are expressed. In fact, it is a *de facto* screen for developmental regulators because the only ones that are not expressed are the Hox and Sox and similar genes that are expressed in narrow windows of development. Having said this, we can narrow it down quite quickly to about 30 macrophage-specific transcription factors that are constitutive and another 20-odd that are inducible. We haven't done much work on anything other than macrophages. In the neutrophils there are four transcription factors that are expressed only in neutrophils and not in macrophages. All four of them have been shown to drive neutrophil differentiation in their own right.

Maizels: When you look at the considerable strain variation that you have found, and if you postulate that one of the differences between two strains is a tissue-specific effect in expressing a particular gene rather than a complete knockout of that particular gene, can you identify that?

Hume: There are a couple of obvious ones. The thing that is very obvious in the BL/6 and DBA, compared with all the other mouse strains, is that they look as if they have already seen interferon (IFN). IRF1 is constitutively up and several of the other IFN-regulated factors are also up. You might predict that the crucial difference between these strains and the others lies somewhere in the IFN pathway. You can do the same for other transcription factors. I think among the mouse strains you can find one that does or doesn't express any one of the genes of interest, or has a very large difference in expression. The one we haven't seen any difference in in the macrophages is PU.1, which is constitutively highly expressed in all the strains. PU.1 knockouts are grossly macrophage deficient. Other than that, everything else shows substantial variation.

Kaufman: One thing that has struck me so far in this meeting is the issue of complexity. There is incredibly complicated alternative splicing. I'm convinced that what you are seeing is real. The question is, is it important? Does it actually do anything? Coming from old school, I think about things in immunology such as class II expression. Class II isn't expressed on any T cell in mice, but in humans and chickens it is present on activated T cells. In horses and frogs it is on all T cells. In axolotols and fish it is not only on T cells and B cells but also on erythrocytes. No one really thinks that this variation means anything. Do you have any idea of how many of the complex things you are describing are going to be important? It is a silly question, I know, but I think it is critical: how do we cleave away what is important to the survival of the organism and what is not?

Hume: In the old days when we started on transcription regulation in mice, it was thought that we should do the 'cat' assay (not standing here for chloramphenicol acetyl transferase, but to throw the mouse in with a cat and see whether it survives). You can never know whether you have actually subjected the animal to the ultimate selective pressure. My famous namesake said that it is never possible to prove the existence of a miracle; you can only disprove all the alternative explanations. You can't prove that something doesn't have a function, but we can prove that it doesn't have a function in most of the things we care to look at. One of the molecules I find interesting is CD4, which mouse macrophages don't express, but rat and human do. Mouse macrophages have exactly the same promoter architecture and they actually do express CD4, but they express a truncated transcript which doesn't splice. This tells us that promoter and splicing events are co-regulated. The macrophages make a truncated CD4 mRNA that doesn't encode anything whereas the T cells make the full-length message off a different transcription start site. Is it important that macrophages make CD4? Are mice worse off because theirs don't? We don't know. With the alternative splicing, the IRAK2 story is a good example where there is an internal promoter that drives a dominant-negative form of IRAC2. If you stop this from being made the inflammatory process continues for longer. This is a clear-cut example of functional alternative splicing. In the c-Fms gene which Christian Schönbach showed, there is a decoy receptor which is membrane anchored but has no kinase domain. Dendritic cell (DC) polarization stops the production of the full-length c-Fms and increases the production of the decoy receptor. This would make the DCs less responsive to CSF-1. This tends to drive them back in the other direction. I don't believe that DCs and macrophages are in any way functionally different in origin, but there is a degree of polarization of function. The CSF-1 receptor alternative splicing looks like it has something to do with this. So we can pluck examples out like this which suggest it is important.

Kaufman: There has to be some way eventually in which we can decide that for the half of the transcripts that have antisense, whether those antisenses are important in the majority of the places, or just a few.

Schönbach: Before we switch to the antisense transcripts I would like to add to David Hume's comments that some alternative splice products are simply the result of a stochastic process. Splice forms generated by splicing errors or somatic mutations may or may not be seen as functionally relevant unless we see a change in cellular behaviour or change from normal to tumour phenotype. Antisense trancripts associated with sense transcripts were thought to be a preferred targets of RNA editing. In 2005 a paper in *Trends in Genetics* (Neemand et al 2005) showed that there was almost no RNA editing among sense-antisense transcripts. From this viewpoint the function of the antisense RNA itself is not obvious. Perhaps the formation of sense-antisense RNA duplexes which may lead to degradation is one aspect of gene expression regulation by antisense RNA.

Hume: The vast majority of the sense/antisense pairs are regulated convergently. Both members of the pair go up together. We have published on an intronic enhancer in the Fims gene which is actually an antisense promoter (Katayama et al 2005). It makes a transcript, but if we over-express the transcript it has no effect at all. Almost all enhancers are promoters. I think a lot of this antisense non-coding RNA transcription is being driven by enhancers and the transcript itself is not important.

Gros: To pick up on the alternative splicing, there clearly is a role for this. For example, we work on a protein (DMT1/Slc11a2) where the terminal exon in the gene is alternatively spliced. This creates two mRNAs. One has an iron-response element which confers regulation by iron of one of the two isoforms. The two RNAs also code for different proteins. The last 23 amino acids are different and these target the protein either to the plasma membrane or to recycling in the endosome. One is regulated at the cell surface by the availability of iron, and functions as an iron transporter. The other functions as an intracellular iron importer translocating transferring iron from acidified endosomes to the cytoplasm. My concern is a bit like yours: there is a lot of it going on, but our ability to associate it with a specific biological process may be limited. This leaves us with a great amount of apparent mRNA complexity, and little understanding of biological relevance.

Goodnow: People have been working on alternative splicing of CD45 for a long time. How functionally important is this? I think it is, but it has been tough to show this.

Britton: The alternative of throwing the mouse to the cat is throwing the mouse to the infection. To take the example of your BL/6 and DBA strains, which seem to be the two closest in their transcriptional response to lipopolysaccharide (LPS); if you infect these two strains with *Mycobacterium tuberculosis*, then the BL/6 mice live twice as long as the DBA strain. The mechanism underlying this isn't understood. This is an example of a gross response in the whole animal to an infection. The complexity has not been analysed for changes in the coding regions of genes, many of which we know are important in functional outcomes. It is an extremely complex system.

Hume: The non-coding RNA question is interesting. It is quite telling that there are almost no mutations by insertion or mutation that have disrupted a functional non-coding RNA, despite extensive mutagenesis of multiple species. *Drosophila* makes non-coding RNAs in large amounts as well. There are many experiments where non-coding RNAs have been disrupted deliberately, and there is no measurable phenotype. So I am not inclined to the view that non-coding RNAs are a crucial part of the control network. What is seen in the cross species comparisons is that the promoters of the non-coding RNAs are actually more highly conserved than the promoters of the protein-coding RNAs, so there is positional equivalence in the genome in that the promoter region is conserved across species, but the RNA itself is not.

Goodnow: The point I wanted to reiterate is this issue of how important it is going to be to create upward references from all these data, for researchers that are taking a top–down approach. If you are taking a bottom–up approach the pathways are easy to trace. But where these kind of data will begin to be filtered is by people coming from the top–down direction, whether this is mapping traits in different populations or looking at cell behaviour. There, I still see this as being a big gap: how do you go from mapping a trait to a particular region, to accessing all this wealth of information about stuff that is going on in that region?

References

Katayama S, Tomaru Y, Kasukawa T et al 2005 Antisense transcription in the mammalian transcriptome. Science 309:1564–1566

Neeman Y, Dahary D, Levanon EY, Sorek R, Eisenberg E 2005 Is there any sense in antisense editing? Trends Genet 21:544–547

Specifying the patterns of immune cell migration

Jason G. Cyster

Howard Hughes Medical Institute and Department of Microbiology and Immunology, University of California, San Francisco, CA 94143, USA

Abstract. Immune system function depends on getting the right cells to the right place at the right time. Inadequate or inappropriate migration of immune cells is involved in many and perhaps all types of immunological disease. Chemokines have been identified as critical guidance factors that help recruit and position cells at each stage of the immune response. Two-photon imaging of intact lymphoid organs has provided evidence of chemotactic migration of lymphocytes in lymphoid organs. Our work on the role of chemokines as organizers of lymphoid tissues will be briefly summarized. Lymphocyte egress from lymphoid organs is necessary for immune surveillance and for effector cell trafficking to sites of inflammation. Sphingosine-1-phosphate (S1P) receptor 1 and the circulatory lipid, S1P, are required for lymphocyte egress from lymphoid organs. We have recently identified CD69 as a regulator of S1P1 and controller of lymphocyte egress. Current molecular understanding of the lymphocyte egress process will be discussed.

2007 Decoding the genomic control of immune reactions. Wiley, Chichester (Novartis Foundation Symposium 281) p 54–64

To permit adaptive immune responses against localized infections, vertebrates have evolved peripheral lymphoid organs (including the spleen, lymph nodes, and Peyer's patches) that filter and concentrate antigens from the nearby tissue and then display them to lymphocytes. The frequency of lymphocytes specific for any given pathogen is low—perhaps 1 per 100 000 lymphocytes—and it is not possible to include all lymphocyte specificities within each lymphoid organ. Instead, naïve lymphocytes travel continuously between different lymphoid organs, surveying for their specific antigen (Fig. 1).

After entering a lymphoid organ from the blood, B cells move to a subcompartment known as the lymphoid follicle while T cells localize in an adjacent area called the T zone (Cyster 2005). Follicles contain follicular dendritic cells (FDCs) that capture filtered antigens and display them in an intact form. T zones contain bone marrow-derived dendritic cells (DCs) that migrate to this compartment from nonlymphoid sites, carrying internalized antigens that they display as peptide–MHC complexes. Subcompartmentalization of the travelling lymphocytes allows

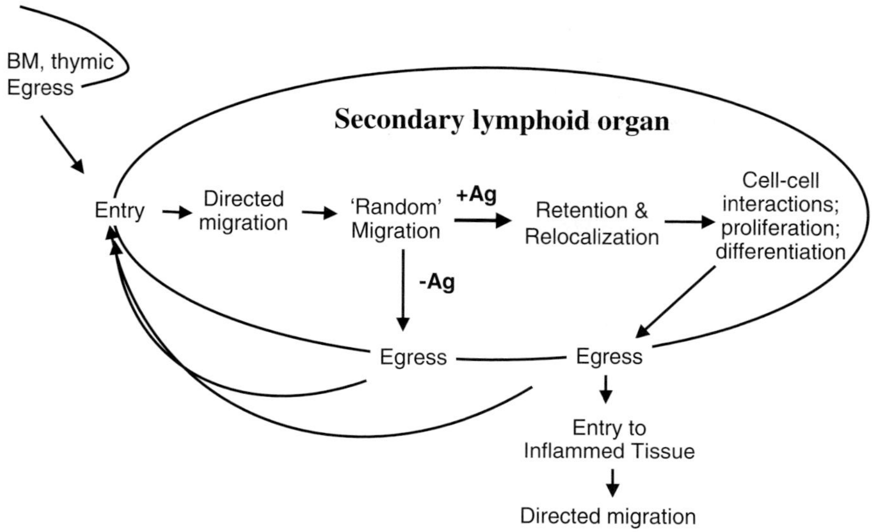

FIG. 1. Patterns of lymphocyte migration during the adaptive immune response. Diagram presents a reductionist view of lymphocyte migration behaviour before and during the early phases of the adaptive immune response. The molecular requirements for each of the indicated processes is under active investigation. Defects in the entry process are a cause of leukocyte adhesion deficiency (LAD) in humans and defects in egress from the bone marrow are a cause of WHIM (Warts, hypogammaglobulinaemia, immunodeficiency and myelokathexis) syndrome. Inhibition of egress from secondary lymphoid organs by the small molecule, FTY720, causes immunosuppression in humans. Ag, antigen; BM, bone marrow.

B cells and T cells to interact efficiently with the appropriate types of antigen-presenting cells. Naïve lymphocytes spend about a day surveying the lymphoid organ for antigen before migrating out of the tissue and returning to circulation. If antigen is encountered, the cells undergo striking changes in migration, stopping within the organ and moving to locations that favour encounters between rare antigen-reactive cells (Fig. 1). Precisely regulated cell movement is therefore essential for immune surveillance and for mounting anti-pathogen immune responses.

Chemotactic cytokines, or chemokines, are small structurally related proteins that are categorized into two principal families, the CXC and CC chemokines (Rot & von Andrian 2004, Cyster 2005). All chemokines signal via seven transmembrane receptors that couple to heterotrimeric G proteins. CXCL13 (BLC) is a CXC chemokine that is made in the B cell follicles and selectively attracts B cells. Using gene targeting in mice, we demonstrated that CXCL13 is necessary for B cell localization within follicles. CXCL13 also guides B cells to body cavities, and CXCL13-deficient mice have compromised body cavity immunity. Two CC chemokines, CCL19 (ELC) and CCL21 (SLC), that are ligands for CCR7 are expressed by stromal cells in the T zone and attract T cells and DCs into this compartment.

Most antibody responses depend on the interaction of antigen-reactive B cells with antigen-specific helper T cells. Early studies hinted that these encounters are favoured by activated B cells and T cells migrating to a common microcompartment within the lymphoid tissue, the outer T zone. We found that antigen-receptor engagement causes B cells to undergo a twofold increase in CCR7 expression and, using B cells lacking or overexpressing CCR7, we showed that this small change was necessary and sufficient to cause B cell relocalization (Reif et al 2002). Antigen-engaged B cells retain responsiveness to CXCL13, and this helps ensure the cells are distributed at the boundary of the follicle and T zone. Thus, a small shift in the balance of responsiveness to chemokines emanating from adjacent zones is sufficient to precisely control cell position.

By carrying out real-time deep-tissue multiphoton microscopy on intact lymph nodes in a collaboration with Mark Miller, Ian Parker and Mike Cahalan (UC Irvine) we showed that antigen-engaged B cells move to the B-T boundary by chemotaxis (Okada et al 2005). Initially the activated B cells showed reduced motility but within one day of antigen-encounter they had increased their motility and they maintained this behaviour while making stable conjugates with helper T cells. Many cognate interactions between B cells and helper T cells lasted 10 to 40 minutes and some interactions persisted for > 1 hour, whereas cells forming non-cognate interactions dissociated in less than 10 minutes.

In addition to guiding cells to lymphoid follicles, CXCL13 and its receptor, CXCR5, also play important roles in lymph node and Peyer's patch development (Ansel & Cyster 2001, Mebius 2003). Organization of germinal centres (GCs), the site of antibody affinity maturation, into 'dark' and 'light' zones depends on the chemokine CXCL12 (SDF1) and CXCR4 as well as CXCL13-CXCR5 (Allen et al 2004). CXCR5 is also required for migration of helper T cells into follicles to support the B cell response. In preliminary work we find that T cells deficient in CXCR5 fail to migrate into B cell follicles following activation. Similar findings have recently been reported by others (Hardtke et al 2005, Junt et al 2005). Transgenic expression of CXCR5 on naive T cells at levels similar to those on activated T cells was sufficient to promote localization of cells near the B-T boundary but was not adequate to direct the cells into B cell follicles (N. Haynes & J.G. Cyster, unpublished observations). These findings again favour the conclusion that changing the balance of responsiveness to chemokines is an important mechanism in controlling cell position.

The mechanisms regulating lymphocyte egress from lymphoid organs have not been well understood. An advance in this area came with the discovery that a small molecule with immunosuppressant properties, FTY720, inhibits lymphocyte emigration (Chiba et al 1998). Two groups established that FTY720 is a substrate for sphingosine kinases, and the phosphorylated form of this molecule is similar to the lysophospholipid sphingosine-1-phosphate (S1P) (Brinkmann et al 2002,

Mandala et al 2002). S1P is present in plasma and body fluids and acts as a signalling molecule by engaging any of five G protein-coupled receptors (S1P1–5). S1P receptors are characterized most prominently by their functions in endothelial cells and for their roles in heart and vascular development.

We found that S1P1 (Edg1) is highly expressed in naïve T and B lymphocytes. Although S1P1-deficiency causes vascular defects and embryonic lethality at mid-gestation, by reconstituting lethally irradiated mice with S1P1-deficient fetal liver cells we found that this receptor is required in the lymphocyte during egress (Matloubian et al 2004). These haematopoietic stem cell chimeric mice lacked T cells in the periphery because mature T cells were unable to exit the thymus. Although B cells were present in peripheral lymphoid organs, they were severely deficient in blood and lymph. Adoptive cell transfer experiments established an intrinsic requirement for S1P1 in T and B cells for lymphoid organ egress. Furthermore, S1P1-dependent chemotactic responsiveness was strongly up-regulated at the step in T cell development prior to exit from the thymus, whereas S1P1 was down-regulated during peripheral lymphocyte activation and this was associated with retention in lymphoid organs.

The model suggested by the above findings was that lymphocyte egress from lymphoid organs was triggered by the higher amounts of S1P in circulatory fluids than in the lymphoid organ (Fig. 2). However, all eukaryotic cell types make S1P

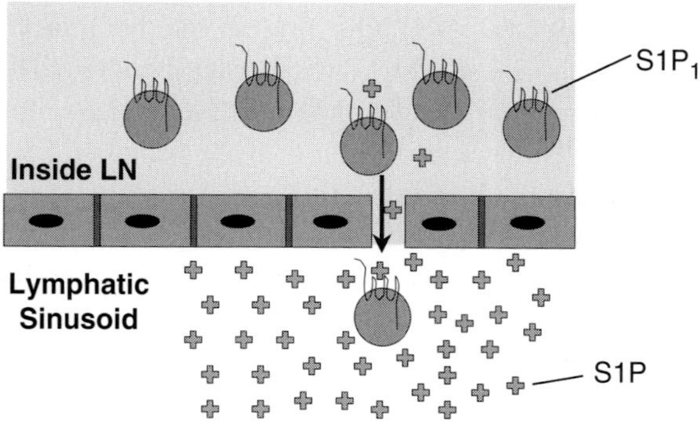

FIG. 2. Model for role of S1P1 in lymphocyte egress from lymphoid organs during homeostasis. Lymphocytes express the receptor S1P1, and S1P levels are high in circulatory fluids such as lymph while being maintained at low levels in lymphoid organs. Lymphocytes are suggested to encounter high S1P levels when they approach egress structures, such as lymphatic sinusoids in lymph nodes and S1P engagement of S1P1 is proposed to transmit an 'egress promoting signal' into the lymphocyte. The nature of this signal is under active investigation. The properties of the endothelial cells (or related stromal cells) at sites of egress is not yet well defined though these structures may contain 'egress portals' (Wei et al 2005). The grey circles represent naïve T or B lymphocytes.

intracellularly and the amounts of S1P present in the interstitial fluids of lymphoid tissues was unclear. We gained insight into the abundance of S1P in lymphoid tissues through experiments to follow-up on the finding that a component of caramel food colorant number III induced lymphopaenia in rodents (Gaunt et al 1977). Treatment of mice with the active component, 2-acetyl-4-tetrahydroxybutylimidazole (THI) caused an inhibition of egress from lymph nodes into lymph (Schwab et al 2005). During earlier studies the effect of THI was found to be overcome by excess dietary vitamin B6 (Sinkeldam et al 1988). Since S1P lyase, the enzyme that terminally degrades S1P, is a vitamin B6-dependent enzyme, we tested whether THI treatment led to changes in lyase activity and S1P levels. THI treatment led to inhibition of S1P lyase activity in thymic extracts and caused S1P abundance in lymphoid tissues to increase more than 100-fold (Schwab et al 2005). These changes were associated with down-modulation of S1P1 from the lymphocyte surface and reduced ability to respond to S1P. These findings indicate that S1P lyase is needed to maintain low S1P levels in lymphoid tissues and they support the model that lymphocytes exit in response to an S1P signal derived from the circulatory fluids or produced locally at exit sites.

Early studies on lymphocyte trafficking demonstrated that egress from lymphoid organs is a regulated process. Hall and Morris found during studies in the sheep that exposure to an inflammatory stimuli caused a transient 'shutdown' in egress of lymphocytes from the draining lymph node into efferent lymph (Hall & Morris 1965). Other studies, including work by Sprent and co-workers, showed that antigen-specific cells could be selectively retained in lymphoid organs for more prolonged periods (Sprent et al 1971). Our finding that lymphocyte expression of S1P1 was critical for egress led us to test whether lymphocyte S1P1 was a physiological egress control point. In initial studies we found that activation of T cell receptor transgenic T cells *in vivo* with their cognate antigen led to 50-fold down-regulation of S1P1 transcript abundance and a loss of S1P responsiveness within one day of antigen exposure (Matloubian et al 2004). By day 3 of exposure, when the activated cells begin egressing as effector cells, S1P1 transcript abundance and S1P responsiveness were restored.

To study the global egress shutdown process, we examined the mechanism of shutdown induced by type I interferons (IFNα/β). Treatment with polyriboinosinic:polyribocytidylic acid (polyI:C), a potent inducer of IFNα/β, led to a shutdown in egress that was dependent on expression of the type I IFNα/β receptor in the lymphocyte. Moreover, lymphocytes from polyI:C treated mice had lost their ability to migrate in response to S1P *in vitro* and had strongly down-regulated surface S1P1 expression (Shiow et al 2006). However, these cells only showed a two–three-fold transcriptional down-regulation of S1P1, leading us to ask whether post-translational regulation of S1P1 was occurring. At this point, the work merged with parallel studies ongoing in the lab to follow-up on observations suggesting an interplay between the early activation antigen, CD69 and S1P1. In particular, we had found that S1P1-deficient T and B cells constitutively expressed S1P1 on the cell surface, while not displaying any other change suggestive of an activated state (Matloubian et al 2004). Moreover, two studies had shown that transgenic overexpression of CD69 in thymocytes led to retention of cells in the thymus (Feng et al 2002, Nakayama et al 2002). However, the interpretation of these studies was complicated by the finding that CD69 deficiency did not affect T cell egress from the thymus (Lauzurica et al 2000). Further indication of an interplay between CD69 and S1P1 came from the finding that S1P1 was identified in a retroviral overexpression screen as a suppressor of surface CD69 expression (Chu et al 2003). Taken together, these observations led us to hypothesize that CD69 may function as a negative regulator of S1P1 and thus of lymphocyte egress. Consistent with this hypothesis, CD69-deficient lymphocytes were poorly retained in lymphoid organs following treatment with polyI:C. In contrast to wild-type lymphocytes, IFNα/β-exposed CD69-deficient lymphocytes retained much of their S1P responsiveness and S1P1 surface expression. Using experiments in cell lines we were able to show that CD69 physically associated with S1P1 and inhibited the function of this receptor in promoting egress (Shiow et al 2006). The mechanism of negative regulation is under further investigation.

The expanding knowledge of the molecular mechanisms specifying immune cell migration obtained from studies in model organisms has provided important insights into a variety of human diseases. Several genetic defects in humans, such as the leukocyte adhesion deficiencies (LAD) I–III, have been found to involve defects in the ability of leukocytes to attach to inflamed endothelium and the understanding of these defects has fit well with the rolling-triggering-adhesion model of leukocyte attachment developed through several decades of research in model systems (Etzioni & Alon 2004). Warts, hypogammaglobulinaemia, immunodeficiency and myelokathexis (WHIM) syndrome was recently identified to be due to activating truncation mutations of CXCR4 (Diaz 2005). Many aspects of this disease are consistent with predictions from studies of CXCR4 function in the mouse. CCR5 deficiency confers resistance to HIV infection but was also recently found in both mouse models and in human studies to lead to increased susceptibility to lethal west Nile virus infection, perhaps due to defective migration of leukocytes into the CNS (Glass et al 2006). Leukocyte recruitment and retention is a necessary process during most immunological diseases, whether acute inflammatory disorders or chronic inflammatory diseases. As well as helping us understand the basis for a variety of immunological diseases, research in this area continues to point towards new therapeutic strategies for regulating unwanted immune responses through control of immune cell migration.

Acknowledgements

I thank Chris Allen, Guy Cinamon, Nicole Haynes, Charles Lo, Mehrdad Matloubian, Takaharu Okada, Susan Schwab and Larry Shiow for work that contributed to the findings summarized in this meeting presentation as well as all the past members of my laboratory for their contributions to laying the foundations for this work. For excellent technical support I thank Jinping An, Olivia Lam and Ying Xu. I also thank Mark Miller, Ian Parker and Mike Cahalan for collaborating on 2-photon microscopy studies and Martin Lipp, Reinhold Forster, and Richard Proia for collaborations involving gene targeted mice. This work was supported by the Howard Hughes Medical Institute and grants from the National Institutes of Health and the David and Lucile Packard Foundation.

References

Allen CD, Ansel KM, Low C et al 2004 Germinal center dark and light zone organization is mediated by CXCR4 and CXCR5. Nat Immunol 5:943–952

Ansel KM, Cyster JG 2001 Chemokines in lymphopoiesis and lymphoid organ development. Curr Opin Immunol 13:172–179

Brinkmann V, Davis MD, Heise CE et al 2002 The immune modulator FTY720 targets sphingosine 1-phosphate receptors. J Biol Chem 277:21453–21457

Chiba K, Yanagawa Y, Masubuchi Y et al 1998 FTY720, a novel immunosuppressant, induces sequestration of circulating mature lymphocytes by acceleration of lymphocyte homing in rats. I. FTY720 selectively decreases the number of circulating mature lymphocytes by acceleration of lymphocyte homing. J Immunol 160:5037–5044

Chu P, Pardo J, Zhao H et al 2003 Systematic identification of regulatory proteins critical for T-cell activation. J Biol 2:21

Cyster JG 2005 Chemokines, sphingosine-1-phosphate, and cell migration in secondary lymphoid organs. Annu Rev Immunol 23:127–159

Diaz GA 2005 CXCR4 mutations in WHIM syndrome: a misguided immune system? Immunol Rev 203:235–243

Etzioni A, Alon R 2004 Leukocyte adhesion deficiency III: a group of integrin activation defects in hematopoietic lineage cells. Curr Opin Allergy Clin Immunol 4:485–490

Feng C, Woodside KJ, Vance BA et al 2002 A potential role for CD69 in thymocyte emigration. Int Immunol 14: 535–544

Gaunt IF, Lloyd AG, Grasso P, Gangolli SD, Butterworth KR 1977 Short-term study in the rat on two caramels produced by variations of the 'ammonia process'. Food Cosmet Toxicol 15:509–521

Glass WG, McDermott DH, Lim JK et al 2006 CCR5 deficiency increases risk of symptomatic West Nile virus infection. J Exp Med 203:35–40

Hall JG, Morris B 1965 The immediate effect of antigens on the cell output of a lymph node. Br J Exp Pathol 46:450–454

Hardtke S, Ohl L, Forster R 2005 Balanced expression of CXCR5 and CCR7 on follicular T helper cells determines their transient positioning to lymph node follicles and is essential for efficient B-cell help. Blood 106:1924–1931

Junt T, Fink K, Forster R et al 2005 CXCR5-dependent seeding of follicular niches by B and Th cells augments antiviral B cell responses. J Immunol 175:7109–7116

Lauzurica P, Sancho D, Torres M et al 2000 Phenotypic and functional characteristics of hematopoietic cell lineages in CD69-deficient mice. Blood 95:2312–2320

Mandala S, Hajdu R, Bergstrom J et al 2002 Alteration of lymphocyte trafficking by sphingosine-1-phosphate receptor agonists. Science 296:346–349

Matloubian M, Lo CG, Cinamon G et al 2004 Lymphocyte egress from thymus and peripheral lymphoid organs is dependent on S1P receptor 1. Nature 427:355–360
Mebius RE 2003 Organogenesis of lymphoid tissues. Nat Rev Immunol 3: 292–303
Nakayama T, Kasprowicz DJ, Yamashita M et al 2002 The generation of mature, single-positive thymocytes in vivo is dysregulated by CD69 blockade or overexpression. J Immunol 168:87–94
Okada T, Miller MJ, Parker I et al 2005 Antigen-engaged B cells undergo chemotaxis toward the T zone and form motile conjugates with helper T cells. PLoS Biology 3:e150
Reif K, Ekland EH, Ohl L et al 2002 Balanced responsiveness to chemoattractants from adjacent zones determines B-cell position. Nature 416:94–99
Rot A, von Andrian UH 2004 Chemokines in innate and adaptive host defense: basic chemokinese grammar for immune cells. Annu Rev Immunol 22:891–928
Schwab SR, Pereira JP, Matloubian M, Xu Y, Huang Y, Cyster JG 2005 Lymphocyte sequestration through S1P lyase inhibition and disruption of S1P gradients. Science 309:1735–1739
Shiow LR, Rosen DB, Brdickova N et al 2006 CD69 acts downstream of interferon-alpha/beta to inhibit S1P1 and lymphocyte egress from lymphoid organs. Nature 440:540–544
Sinkeldam EJ, de Groot AP, van den Berg H, Chappel CI 1988 The effect of pyridoxine on the number of lymphocytes in the blood of rats fed caramel colour (III). Food Chem Toxicol 26:195–203
Sprent J, Miller JFAP, Mitchell GF 1971 Antigen-induced selective recruitment of circulating lymphocytes. Cell Immunol 2:171–181
Wei SH, Rosen H, Matheu MP et al 2005 Sphingosine 1-phosphate type 1 receptor agonism inhibits transendothelial migration of medullary T cells to lymphatic sinuses. Nat Immunol 6:1228–1235

DISCUSSION

Mackay: I was interested in the modest change in CCR5 and the profound effect it had on CCR7. With CCR5 we see enormous interindividual differences. Around this table we would have people who are almost null and others will express high levels. This associates with a lot of polymorphisms in the CCR5 promoter and other regions. What is happening with CCR5 is that it is almost like an immune response gene that has been subjected to polymorphisms, with a need to either soup up or damp down immune responses. The levels of that chemokine receptor are probably somewhat similar to the effect you are seeing with your chemokine receptor having such profound effects on cell migration.

Cyster: So your point is that even small effects might have phenotypic consequences, and the standard variation in the population is enough to cause huge phenotypic variation?

Mackay: Yes, and CCR5 is particularly associated with inflammatory responses.

Goodnow: Is it known to be a receptor for a particular human pathogen?

Mackay: In general, it is a Th1-associated receptor.

Lam: You mentioned that B cell receptor (BCR) engagement upregulates the CCR7 expression, but the level of expression of CXCR5 remains the same on B cells. Is this retained level required for the B cells to migrate to the T cell zone?

Cyster: I should clarify one thing. CXCR5 will change later in the response. As they differentiate from B cells to plasma cells they completely down-regulate CXCR5. If they don't have CXCR5 to begin with they can't get into the follicle. If you take it away when they are engaged by antigen they still don't move uniformly into the T zone, so it appears that the receptor contributes to their distribution along the B/T boundary region but it is not the only thing that is limiting their access to the T zone. Additional factors are helping to control fine position.

Lam: So no one has done inducible knockout of CXCR5 at the point of migration into the T zone.

Cyster: That is right.

Vinuesa: We tend to equate lymphadenopathy for example with lymphoproliferation. I was intrigued to see that CD69 up-regulation could block egress from the lymph nodes. Quite a few of our autoimmune mouse models have lymphadenopathy, but do not have any signs of increased T cell or B cell proliferation. To what extent do you think that this inhibition of egress due to CD69 up-regulation could be a determinant for lymphadenopathy and not a consequence of lymphoproliferation?

Cyster: Physiologically, this process can act rapidly. Just as the effects on egress were observed a long time ago, the entry process was observed a long time ago to be augmented as well. A huge increase in influx takes place through a variety of changes in the vasculature. Some is very rapid because of vasodilation, and some involves growth or further high endothelial venule development. Both entry increases and exit decreases, which will contribute to lymphadenopathy. We think the CD69 effect is rather transient. However, in T cells that have been stimulated by the T cell receptor, S1P1 can be markedly transcriptionally down-regulated. This will cause retention also. There may be additional extrinsic mechanisms that can also regulate the egress process outside the lymphocyte. The FTY720 molecule that can artificially inhibit this process may at least in part be acting extrinsically.

Malissen: How does your CD69/S1P1 story fit with the thymus? Essentially, the positive selection with the T cell receptor (TCR) is supposed to induce CD69.

Cyster: Most newly developed single positive thymocytes have CD69 and then as they mature they down-regulate it. At least part of this down-regulation is now known to be S1P1 dependent at the protein level. When we purify naïve lymphocytes from S1P1-deficient mice they have CD69 constitutively. The way CD69 is suppressed in the cell seems to be by complexing with S1P1. We presume S1P1 is more abundant in the naïve cell and keeps CD69 from the surface. If we knock S1P1 out CD69 comes to the surface. We still don't know how this is playing out in the thymus.

Malissen: Is it the ectodomain of CD69 that interacts with S1P1?

Cyster: It is looking complicated. We don't have a full handle on this.

Gros: On your slides you draw S1P as a soluble ligand, but it is a phospholipid. How much of it can be in the membrane? Perhaps the concentration in the membrane can act as some sort of homing focus.

Cyster: There is very little of it in the membrane because it only has a single hydrocarbon tail. This isn't sufficient to restrain it in the membrane. Very little of it is bioavailable versus bound to carriers. There is indirect evidence for this in that the K_d of the receptor is nanomolar, while the amounts in circulation are micromolar, yet the receptor is not saturated. The major carriers turn out to be HDL and albumin. How you control the bioavailable fraction is not clear.

Turner: When you get resolution of infection, quite often you can find long-lived memory cells in the lymphoid tissues. One observation we made is that these cells have up-regulated CD69. Is a similar sort of mechanism inhibiting egress from tissues after the resolution of infection?

Cyster: That is a good question. We haven't had a chance to look extensively at this. The closest thing we did early on is to look at DCs that were homing to lymph nodes. We saw a rather limited effect of FTY720 treatment there. It may be a fairly specialized process. CCR7, for example, is important for getting memory cells into peripheral lymphatics, but we don't know to what degree S1P1 will also be required.

Goodnow: So you think in this system S1P might not be an important attractant for cell egress from inflamed solid organs?

Cyster: At this point we lack data indicating a critical role, but most experiments that can be done at this level currently are with small molecules. Then you never know what the intrinisic effects are. The knockout is lethal. What is clear is that peripheral lymphatics have lots of CCR7 ligand, and there are a couple of studies showing the effector cells that will return from the inflamed site have to have CCR7. There seems to be another cue (the CCR7 ligand, CCL21) acting as a critical player in that case.

Kaufman: You have shown us that the B cells go to the B cell follicles because they follow some chemokines, and T cells go to the paracortex because they follow others. Who sets down the chemokines?

Cyster: That's a developmental biology question. There is a reinforcement of the compartments in the stromal development by the lymphocytes. B cells induce FDC maturation and increased expression of the B cell chemoattractant factor, so there is a positive feedback. They don't dictate the initial pattern. The guess is that this will have been set by very early inductive events, such as homeobox-type patterning.

Vinuesa: What is your feeling about the frantic migration of the B cell once it has already attached to a T cell? Is this random, because they still haven't found the stop signal, or are they still seeking the elusive ménage a trois with the DC?

Cyster: We haven't seen an indication that they are going to stop. They themselves saw antigen earlier and this will probably relate to how much they continue to see antigen. The notion they might also benefit from encountering a DC is a possibility. Whether they benefit from moving around and potentially exchanging T cell partners in a diverse repertoire is something one could speculate about.

Hume: S1P1 is highly expressed on macrophages but is completely ablated by TLR signalling.

Cyster: In the B cell we see that TLR signalling drives S1P1 expression down. Why macrophages are using it at the outset is less clear.

Human monogenic disorders that confer predisposition to specific infections

Capucine Picard*†, Laurent Abel* and Jean-Laurent Casanova*‡

**Laboratoire de Génétique Humaine des Maladies Infectieuses, Université de Paris René Descartes-INSERM U550, Faculté de Médecine Necker, 156 Rue de Vaugirard, 75015 Paris, France, † Centre d'étude des déficits immunitaires, Hôpital Necker-Enfants Malades, 149 Rue de Sèvres, 75015 Paris, France and ‡ Unité d'Immunologie et d'Hématologie Pédiatriques, Hôpital Necker-Enfants Malades, 149 Rue de Sèvres, 75015 Paris, France*

Abstract. Human monogenic disorders affecting immunity to infection are increasingly recognized. Most are associated with primary immunodeficiencies, which typically confer predisposition to multiple infectious diseases. We review here the known, atypical monogenic disorders that confer a narrow vulnerability to infection with specific microorganisms in otherwise healthy individuals. These 'experiments of nature' have important immunological and clinical implications.

2007 Decoding the genomic control of immune reactions. Wiley, Chichester (Novartis Foundation Symposium 281) p 65–78

Only a minority of infected individuals develop clinical disease and human genes play an important role in determining susceptibility or resistance (Casanova & Abel 2005). Predisposition or resistance to infectious disease may follow a simple or complex pattern of inheritance. Inborn errors of immunity affecting a single gene (described as Mendelian) are often referred to as primary immunodeficiencies (PIDs). PIDs include over 200 known clinical syndromes, at least 100 of which have a well defined molecular genetic basis (Notarangelo et al 2004). PIDs are typically associated with multiple and recurrent infections caused by weakly virulent (opportunist) microorganisms. However, severe opportunistic infectious diseases, often recurrent, have also been described in otherwise healthy individuals. The high frequency of familial forms, parental consanguinity, or both, suggested that these 'idiopathic' infections reflected a Mendelian predisposition. This intriguing group of 'non-conventional' PIDs (Casanova et al 2005), characterized by a narrow spectrum of specific infections, limited to one microbial genus or species, is small but expanding (Casanova et al 2002, 2005, Casanova & Abel 2004b, 2005). The specificity of predisposition to infections is not absolute, but these disorders are readily distinguished from other, 'conventional' PIDs on purely clinical grounds. The discovery of similar phenotypes and their corresponding genotypes in mice, with natural mutations conferring vulnerability to specific infections (e.g. *Mx*, *Bcg*, *Lps*, *Cmv*), has opened up new, fundamental avenues of research in molecular immunology (Casanova et al 2002). We review here the well documented monogenic disorders known to confer susceptibility to specific infections in humans, focusing on immunological implications.

Predisposition to invasive *Neisseria* disease

Molecular defects in the three pathways of complement activation (classical, alternative, lectin-mediated) are associated with various autoimmune, allergic and infectious phenotypes (Sullivan & Winkelstein 1999). Two types of complement deficiency are responsible for a selective predisposition to recurrent invasive infections caused by *Neisseria* species, and *N. meningitidis* (common and uncommon serotypes) in particular (Fijen et al 1999). These deficiencies were the first examples of Mendelian susceptibility to a specific infection to be elucidated at the molecular level. The first, described in 1974, affects the terminal components of complement (C5b–C9), which form the membrane attack complex (MAC) downstream from the three pathways of complement activation (Sullivan & Winkelstein 1999). Autosomal recessive mutations in *C5*, *C6*, *C7*, *C8A*, *C8B*, *C8G* and *C9* have been identified since 1993 in more than 250 patients (Sullivan & Winkelstein 1999, Wurzner 2003). The second cause of predisposition to *Neisseria* infection, identified in 1982, is properdin deficiency (Sullivan & Winkelstein 1999). Properdin is a glycoprotein that acts as a positive regulator of the alternative pathway and, indirectly ensures the optimal functioning of the complement cascade initiated by the classical or lectin pathways (Sullivan & Winkelstein 1999). The *PFC* gene encoding properdin is located on the X chromosome and hemizygous mutations have been reported in more than 100 male patients since 1995 (Sullivan & Winkelstein 1999). Interindividual differences in susceptibility to meningococcal disease have been observed in properdin-deficient patients, possibly due to other genetic factors (Emonts et al 2003). For example, a recent study of a Danish kindred indicated that mannose-binding lectin (MBL) deficiency, affecting the third pathway of complement activation, may be a supplementary risk factor in properdin-deficient male patients (Bathum et al 2006). Isolated MBL deficiency, a common trait in human populations, has no major impact on predisposition to infection (Casanova & Abel 2004a). These 'experiments of nature' demonstrate the requirement for both complement pathway stabilisation by properdin and the integrity of the MAC for the destruction of invading *Neisseria*, in natural conditions of infection (Casanova & Abel 2004b), and their redundancy for protective immunity to other microbes.

Mendelian susceptibility to mycobaterial disease

Mendelian susceptibility to mycobaterial disease (MSMD) was probably first clinically described in 1951 (Casanova & Abel 2002). Patients with MSMD are highly susceptible to weakly virulent mycobacteria (environmental mycobacteria and BCG vaccines), but are apparently resistant to most other infectious agents, with the exception of *Salmonella* (Fieschi et al 2003, Dorman et al 2004). Since 1996, disease-causing mutations have been found in five autosomal genes (*IFNGR1*, *IFNGR2*, *STAT1*, *IL12B* and *IL12RB1*), involved in interleukin (IL)12/23-dependent, interferon (IFN)γ-mediated immunity (Casanova & Abel 2002, Casanova et al 2005). Extensive allelic heterogeneity at these loci accounts for the existence of 12 distinct genetic disorders, which have been diagnosed in more 300 patients worldwide (Fieschi et al 2003, Dorman et al 2004, Vogt et al 2005, Filipe Santos et al 2006) (Chapgier et al 2006). Recent discoveries include the description of the first *IFNGR2* mutation associated with the surface expression of non-functional IFNγ R2 chains (Vogt et al 2005). This missense mutation creates a novel consensus site for *N*-glycosylation (Vogt et al 2005). The resulting polysaccharide is pathogenic, and the defect can be complemented *in vitro* by treatment with inhibitors of glycosylation and glycosidases. Interestingly, up to 1.4% of human missense mutations in proteins transported in the secretory pathway are predicted to lead to a gain-of-glycosylation. Two new mutations of *STAT1*, affecting the DNA-binding domain of Stat1, have also been shown to cause partial dominant Stat1 deficiency (Chapgier et al 2006). Together with the previously described loss-of-phosphorylation mutation in *STAT1* (Dupuis et al 2001), these mutations define a new class of mutant alleles, deleterious for two phenotypes, but dominant for one and recessive for the other (*domissive* alleles). Finally, MSMD-causing *NEMO* mutations in three unrelated kindreds defined the first genetic aetiology of X-linked recessive MSMD (Frucht & Holland 1996, Filipe-Santos et al 2006). These missense mutations affect two nearby residues in the leucine zipper domain and impair the CD40-triggered induction of IL12 production by monocyte-derived cells upon stimulation by CD40L-expressing T cells, accounting for the observed susceptibility to mycobacteria (Filipe-Santos et al 2006). Conversely, these mutations do not affect NF-κB activation in response to classical activators such as lipopolysaccharide (LPS), tumour necrosis factor (TNF)α and IL1β, accounting for the patients' resistance to other infections (Filipe-Santos et al 2006). Studies of MSMD have shown that the IL12/23–IFNγ circuit is crucial for protective immunity against mycobacteria and *Salmonella* but, paradoxically, redundant against most other microorganisms (Fieschi et al 2003, Dorman et al 2004).

Predisposition to *Streptoccocus pneumoniae* infection

The pathogenesis of invasive pneumococcal disease is unclear, as the causal microorganism is innocuous in most infected individuals (Picard et al 2003b). Rare patients with congenital asplenia, complement deficiency or antibody deficiency, are vulnerable to multiple encapsulated bacteria, including *Streptococcus pneumoniae.* An apparently more selective predisposition to pneumococci was documented in inherited IL1 receptor-associated kinase 4 (IRAK4) deficiency, an autosomal recessive disorder first described in 2003 in three unrelated patients (Picard et al 2003a). Seventeen other patients have since been identified (Medvedev et al 2003, Currie et al 2004, Enders et al 2004, Chapel et al 2005, Yang et al 2005, Cardenes et al 2006, Takada et al 2006, J.-L. Casanova, unpublished data). Clinically, IRAK-4-deficient patients present recurrent infections caused by pyogenic bacteria, including *S. pneumoniae* in particular, with weak or delayed inflammatory responses. Only three of the identified patients had invasive disease caused by Gram-negative bacteria (Medvedev et al 2003, Chapel et al 2005). IRAK4 deficiency is a life-threatening disease, with eight deaths (caused by pneumococcus infection in six cases) among 20 patients (Medvedev et al 2003, Currie et al 2004, Enders et al 2004, Yang et al 2005, Chapel et al 2005, Cardenes et al 2006, Takada et al 2006, J.-L. Casanova, unpublished data), but the global trend shows a clinical improvement with age (Medvedev et al 2003, Chapel et al 2005). The patients' blood cells fail to produce pro-inflammatory cytokines upon stimulation with Toll-like receptor (TLR) agonists, IL1β and IL18. Recent discoveries include the lack of IFNα/β and λ induction via TLR7, TLR8 and TLR9 in IRAK4-deficient blood cells (Yang et al 2005). In contrast, IFNα/β and λ are induced normally via TLR3 and TLR4, and in response to most viruses. Thus, IRAK4-deficient patients may control viral infections via the TLR3- and TLR4-dependent and/or TLR-independent production of anti-viral IFNs. The TLR7-, TLR8-, and TLR9-dependent induction of IFNα/β and λ is nonetheless strictly IRAK4-dependent and, paradoxically, redundant for protective immunity to most viruses in humans (Yang et al 2005). Intriguingly, the TLR and IL1R signalling pathways are important for immune responses to pneumococci, but seem to be redundant for protective immunity to most other pathogens.

X-linked predisposition to Epstein-Barr virus infection

X-linked recessive lymphoproliferative disease (XLP), first characterized clinically in 1975, is associated with susceptibility to B cell-tropic Epstein-Barr virus (EBV) (Purtilo et al 1975). The clinical manifestations are heterogeneous but reflect impaired host control of EBV-infected B cells, and include fulminant infectious mononucleosis with haemophagocytosis (excessive T cell-mediated macrophage activation in response to the primary infection of B cells with EBV), B cell lymphoproliferative disorders and lymphomas (possibly resulting from latent EBV infection), and hypo-gammaglobulinaemia (which also involves B cells, but has an unknown underlying mechanism) (Nichols et al 2005b). XLP is caused by hemizygous mutations in *SH2D1A* (Sayos et al 1998), and more than 300 affected male patients have been reported since 1998. This gene encodes a small cytoplasmic adapter known to act as a signalling lymphocytic activation molecule (SLAM)-associated protein (SAP) (Sayos et al 1998). SAP is produced in T and NK cells, and regulates the function of SLAM family receptors, such as 2B4, the signalling pathway of which is impaired in XLP patients (Latour & Veillette 2004). SAP-deficient NK and $CD8^+$ cytotoxic T cells are present but do not efficiently kill EBV-transformed B cells, probably due to impaired 2B4 signalling following stimulation with the natural ligand, CD48, on EBV-infected cells (Parolini et al 2000, Sharifi et al 2004, Dupre et al 2005). XLP $CD4^+$ T helper cells have also been shown to be deficient, with impaired induction of IL10 (upon stimulation by anti-CD3 antibody) and ICOS (upon stimulation by PHA plus IL2)—a potent inducer of IL10 (Ma et al 2005). The cause of the IL10 deficiency is unclear, but this deficiency may contribute to some of the B cell abnormalities seen in XLP patients, such as the lack of detectable $CD27^+$ switched memory B cells, which was documented before EBV infection in some patients (Ma et al 2006). Finally, the development of NKT cells is severely impaired in XLP patients (Nichols et al 2005a, Pasquier et al 2005). This may be related to the recent observation of an EBV-driven lymphoma in a patient with a hitherto unknown genetic defect and a combined NK and NKT cell deficiency (Eidenschenk et al 2006). However, the cellular and molecular mechanism by which SAP mutations are involved in the pathogenesis of EBV-driven XLP remains unclear, as does the reason for the apparent resistance of the patients to other pathogens, including other B cell-tropic viruses.

Predisposition to papillomaviruses: epidermodysplasia verruciformis

Epidermodysplasia verruciformis (EV) is characterized by selective susceptibility to infections caused by skin-tropic oncogenic papillomaviruses (HPVs) of the B1 group, leading to an increase in the risk of epithelial skin cancer. It was first described clinically in 1922 (Jablonska & Majewski 1994). Autosomal recessive mutations in two adjacent genes, *EVER1* and *EVER2*, were described in 2002 in three and two kindreds, respectively (Ramoz et al 2002). One patient with compound heterozygous mutations in *EVER1* and another with a homozygous mutation in *EVER2* were recently described (Tate et al 2004, Sun et al 2005). *EVER* genes belong to the transmembrane channel-like (TMC) family of genes, encoding proteins with eight membrane-spanning domains, thought to act as ion channels, transporters, or their associated modifiers (Kurima et al 2003, Keresztes et al 2003). The TMC family was first discovered by positional cloning of human inherited deafness (Kurima et al 2002). *EVER1/TMC6* and *EVER2/TMC8* are normally expressed in the endoplasmic reticulum of keratinocytes (Ramoz et al 2002). Despite recent progress, many questions about the function of EVER1/TMC6 and EVER2/TMC8 and the role of their mutant alleles in the pathogenesis of EV remain unresolved. Patients with severe combined immunodeficiency (SCID) caused by γc and JAK3 deficiencies, but not SCID patients bearing other molecular defects, were recently shown to develop an EV-like clinical syndrome, despite successful haematopoietic stem cell transplantation (Laffort et al 2004). This probably reflects impaired expression of the two genes in keratinocytes, and possibly the impaired binding of their products with those of *EVER1* and *EVER2* (Laffort et al 2004). The *in vitro* study of these genes in keratinocytes and the identification of new genetic aetiologies of EV should provide insight into protective immunity to papillomaviruses and the pathogenesis of papillomavirus diseases in selected patients.

Conclusions

Five Mendelian traits conferring susceptibility to a single, weakly pathogenic pathogen have been elucidated at the molecular level. Two studies identified new genes involved in host defence (XLP, EV), whereas the other three provided insight into the function of known genes in humans (MSMD, MAC/PFC, invasive pneumococcal disease). Many other similar clinical syndromes, such as isolated chronic mucocutaneous candidiasis (Lilic 2002), herpes simplex encephalitis (Dupuis et al 2003), and HHV8-driven idiopathic Kaposi's sarcoma (Guttman-Yassky et al 2004), are probably also Mendelian, but molecular genetic investigations are required to confirm this hypothesis. More common infectious diseases, caused by more virulent micro-organisms, may also result from Mendelian disorders, as illustrated by the recent demonstration that tuberculosis (TB) is a Mendelian disorder, at least in some children (Alcais et al 2005). These 'experiments of nature' demonstrate the existence of specific interactions between certain immune genes and pathogenic microbes. In natural conditions of immunity and infection—the hallmark of the human model (Casanova & Abel 2002, 2004b)—these genes appear to be non-redundant for protective immunity to specific pathogens. The mutations reviewed here have provided important insight into the functioning of the immune system, as they have revealed ecologically relevant and evolutionary selected specific interactions between certain human genes and microbial pathogens.

Acknowledgements

We would like to thank all the members of our laboratory for helpful discussions and critical reading of this manuscript, and our collaborators worldwide for their trust and patience.

Needless to say, we are much indebted to the patients and their families. This work was supported by the Fondation BNP-Paribas, the Fondation Schlumberger, and EU grant number QLK2-CT-2002-00846. JLC is an International Scholar of the Howard Hughes Medical Institute.

References

Alcais A, Fieschi C, Abel L, Casanova JL 2005 Tuberculosis in children and adults: two distinct genetic diseases. J Exp Med 202:1617–1621

Bathum L, Hansen H, Teisner B et al 2006 Association between combined properdin and mannose-binding lectin deficiency and infection with Neisseria meningitidis. Mol Immunol 43:473–479

Cardenes M, von Bernuth H, Garcia-Saavedra A et al 2006 Clinical and laboratory observations autosomal recessive interkeukin receptor associated kinase 4 deficiency in fourth-degree relatives. J Pediatr 148:549–551

Casanova JL, Abel L 2002 Genetic dissection of immunity to mycobacteria: the human model. Annu Rev Immunol 20:581–620

Casanova JL, Abel L 2004a Human mannose-binding lectin in immunity: friend, foe, or both? J Exp Med 199:1295–1299

Casanova JL, Abel L 2004b The human model: a genetic dissection of immunity to infection in natural conditions. Nat Rev Immunol 4:55–66

Casanova JL, Abel L 2005 Inborn errors of immunity to infection: the rule rather than the exception. J Exp Med 202:197–201

Casanova JL, Schurr E, Abel L, Skamene E 2002 Forward genetics of infectious diseases: immunological impact. Trends Immunol 23:469–472

Casanova JL, Fieschi C, Bustamante J et al 2005 From idiopathic infectious diseases to novel primary immunodeficiencies. J Allergy Clin Immunol 116:426–430

Chapel H, Puel A, von Bernuth H, Picard C, Casanova JL 2005 Shigella sonnei meningitis due to interleukin-1 receptor-associated kinase-4 deficiency: first association with a primary immune deficiency. Clin Infect Dis 40:1227–1231

Chapgier A, Boisson-Dupuis S, Jouanguy E et al 2006 Novel STAT1 alleles in otherwise healthy patients with mycobacterial disease. PLoS Genet 2:e131

Currie AJ, Davidson DJ, Reid GS et al 2004 Primary immunodeficiency to pneumococcal infection due to a defect in Toll-like receptor signaling. J Pediatr 144:512–518

Dorman SE, Picard C, Lammas D et al 2004 Clinical features of dominant and recessive interferon gamma receptor 1 deficiencies. Lancet 364:2113–2121

Dupre L, Andolfi G, Tangye SG et al 2005 SAP controls the cytolytic activity of CD8+ T cells against EBV-infected cells. Blood 105:4383–4389

Dupuis S, Dargemont C, Fieschi C et al 2001 Impairment of mycobacterial but not viral immunity by a germline human STAT1 mutation. Science 293:300–303

Dupuis S, Jouanguy E, Al-Hajjar S et al 2003 Impaired response to interferon-alpha/beta and lethal viral disease in human STAT1 deficiency. Nat Genet 33:388–391

Eidenschenk C, Dunne J, Jouanguy E et al 2006 A Novel primary immunodeficiency with specific natural-killer cell deficiency maps to the centromeric region of chromosome 8. Am J Hum Genet 78:721–727

Emonts M, Hazelzet JA, de Groot R, Hermans PW 2003 Host genetic determinants of Neisseria meningitidis infections. Lancet Infect Dis 3:565–577

Enders A, Pannicke U, Berner R et al 2004 Two siblings with lethal pneumococcal meningitis in a family with a mutation in Interleukin-1 receptor-associated kinase 4. J Pediatr 145: 698–700

Fieschi C, Dupuis S, Catherinot E et al 2003 Low penetrance, broad resistance, and favorable outcome of interleukin 12 receptor beta1 deficiency: medical and immunological implications. J Exp Med 197:527–535

Fijen CA, Kuijper EJ, te Bulte MT, Daha MR, Dankert J 1999 Assessment of complement deficiency in patients with meningococcal disease in The Netherlands. Clin Infect Dis 28:98–105

Filipe-Santos O, Bustamante J, Haverkamp MH et al 2006 X-linked susceptibility to mycobacteria is caused by mutations in NEMO impairing CD40-dependent IL-12 production. J Exp Med 203:1745–1759

Frucht DM, Holland SM 1996 Defective monocyte costimulation for IFN-gamma production in familial disseminated Mycobacterium avium complex infection: abnormal IL-12 regulation. J Immunol 157:411–416

Guttman-Yassky E, Cohen A, Kra-Oz Z et al 2004 Familial clustering of classic Kaposi sarcoma. J Infect Dis 189:2023–2026

Jablonska S, Majewski S 1994 Epidermodysplasia verruciformis: immunological and clinical aspects. Curr Top Microbiol Immunol 186:157–175

Keresztes G, Mutai H, Heller S 2003 TMC and EVER genes belong to a larger novel family, the TMC gene family encoding transmembrane proteins. BMC Genomics 4:24

Kurima K, Peters LM, Yang Y et al 2002 Dominant and recessive deafness caused by mutations of a novel gene, TMC1, required for cochlear hair-cell function. Nat Genet 30:277–284

Kurima K, Yang Y, Sorber K, Griffith AJ 2003 Characterization of the transmembrane channel-like (TMC) gene family: functional clues from hearing loss and epidermodysplasia verruciformis. Genomics 82:300–308

Laffort C, Le Deist F, Favre M et al 2004 Severe cutaneous papillomavirus disease after haemopoietic stem-cell transplantation in patients with severe combined immune deficiency caused by common gamma cytokine receptor subunit or JAK-3 deficiency. Lancet 363:2051–2054

Latour S, Veillette A 2004 The SAP family of adaptors in immune regulation. Semin Immunol 16:409–419

Lilic D 2002 New perspectives on the immunology of chronic mucocutaneous candidiasis. Curr Opin Infect Dis 15:143–147

Ma CS, Hare NJ, Nichols KE et al 2005 Impaired humoral immunity in X-linked lymphoproliferative disease is associated with defective IL-10 production by CD4+ T cells. J Clin Invest 115:1049–1059

Ma CS, Pittaluga S, Avery DT et al 2006 Selective generation of functional somatically mutated IgMCD27, but not Ig isotype-switched, memory B cells in X-linked lymphoproliferative disease. J Clin Invest 116:322–333

Medvedev AE, Lentschat A, Kuhns DB et al 2003 Distinct mutations in IRAK-4 confer hyporesponsiveness to lipopolysaccharide and interleukin-1 in a patient with recurrent bacterial infections. J Exp Med 198:521–531

Nichols KE, Hom J, Gong SY et al 2005a Regulation of NKT cell development by SAP, the protein defective in XLP. Nat Med 11:340–345

Nichols KE, Ma CS, Cannons JL, Schwartzberg PL, Tangye SG 2005b Molecular and cellular pathogenesis of X-linked lymphoproliferative disease. Immunol Rev 203:180–199

Notarangelo L, Casanova JL, Fischer A et al 2004 Primary immunodeficiency diseases: an update. J Allergy Clin Immunol 114:677–687

Parolini S, Bottino C, Falco M et al 2000 X-linked lymphoproliferative disease. 2B4 molecules displaying inhibitory rather than activating function are responsible for the inability of natural killer cells to kill Epstein-Barr virus-infected cells. J Exp Med 192:337–346

Pasquier B, Yin L, Fondaneche MC et al 2005 Defective NKT cell development in mice and humans lacking the adapter SAP, the X-linked lymphoproliferative syndrome gene product. J Exp Med 201:695–701

Picard C, Puel A, Bonnet M et al 2003a Pyogenic bacterial infections in humans with IRAK-4 deficiency. Science 299:2076–2079

Picard C, Puel A, Bustamante J, Ku CL, Casanova JL 2003b Primary immunodeficiencies associated with pneumococcal disease. Curr Opin Allergy Clin Immunol 3:451–459

Purtilo DT, Cassel CK, Yang JP, Harper R 1975 X-linked recessive progressive combined variable immunodeficiency (Duncan's disease). Lancet 1:935–940

Ramoz N, Rueda LA, Bouadjar B et al 2002 Mutations in two adjacent novel genes are associated with epidermodysplasia verruciformis. Nat Genet 32:579–581

Sayos J, Wu C, Morra M et al 1998 The X-linked lymphoproliferative-disease gene product SAP regulates signals induced through the co-receptor SLAM. Nature 395:462–469

Sharifi R, Sinclair JC, Gilmour KC et al 2004 SAP mediates specific cytotoxic T-cell functions in X-linked lymphoproliferative disease. Blood 103:3821–3827

Sullivan KE, Winkelstein JA 1999 Genetically determined deficiencies of the complement system. In: Ochs H, Edvard Smith CI, Puck JM (eds) Primary immunodefiency diseases: a molecular and genetic approach, Oxford University Press, p 397–418

Sun XK, Chen JF, Xu AE 2005 A homozygous nonsense mutation in the EVER2 gene leads to epidermodysplasia verruciformis. Clin Exp Dermatol 30:573–574

Takada H, Yoshikawa H, Imaizumi M et al 2006 Delayed separation of umbilical cord in two siblings with interleukin-1 receptor associated kinase 4 deficiency: Rapid screening by flow cytometer. J Pediatr 148:546–548

Tate G, Suzuki T, Kishimoto K, Mitsuya T 2004 Novel mutations of EVER1/TMC6 gene in a Japanese patient with epidermodysplasia verruciformis. J Hum Genet 49:223–225

Vogt G, Chapgier A, Yang K et al 2005 Gains of glycosylation comprise an unexpectedly large group of pathogenic mutations. Nat Genet 37:692–700

Wurzner R 2003 Deficiencies of the complement MAC II gene cluster (C6, C7, C9): is subtotal C6 deficiency of particular evolutionary benefit? Clin Exp Immunol 133:156–159

Yang K, Puel A, Zhang S et al 2005 Human TLR-7-, -8-, and -9-mediated induction of IFN-alpha/beta and -lambda Is IRAK-4 dependent and redundant for protective immunity to viruses. Immunity 23:465–478

DISCUSSION

Goodnow: The impression I get from what you have talked about, is that this might be the tip of the iceberg: as you get to progressively more subtle defects in genes that we know of as Mendelian, the phenotypes will start to look clinically more and more like run-of-the-mill kinds of common diseases.

Casanova: The phenotype I referred to as TB is the clinical phenotype of paediatric TB, typically a primary infection that disseminates to several tissues. Whether it is the tip of the iceberg or not, whether this and other mycobacterial phenotypes are also determines by single-gene lesions, is an open and important question. Are polygenic diseases polygenic because they are a sum of Mendelian diseases, with each individual suffering from a given Mendelian defect? Or are they truly polygenic in individual patients? For the field of infectious diseases in humans like for other fields in human genetics, this is a key question. I favour the idea that there is a continuum, at the population level, between the two forms of genetic determinism.

Goodnow: In the mice, when we are generating equivalent kinds of missense mutations that have a Mendelian trait, one of the things we are learning is how easy it is for a simple Mendelian trait with incomplete penetrance and a bit of variability to look not Mendelian.

Casanova: I agree. There is low penetrance for a number of the diseases I described. There is no such thing as 'Mendelian genetics' because there is no single-gene organism. I did not mean to imply that there are no modifier genes. Once the disease-causing genes are identified comes the time to search for modifier genes.

Wakeland: That is a fairly major point. What you are saying is that you could have a major allele that is leading to a susceptibility, but that this is being modified by a whole variety of additional genes, putting it back into a polygenic phenotype with a major gene.

Morahan: I'd like to share some data we have. We have studied patients with either TB or leprosy from India in collaboration with Prof Narinder Mehra. We see strong association of these diseases with particular *IL12B* genotypes in this adult population. One particular genotype is uncommon in Europeans, but accounts for something like 20% of controls in India, which is suggestive evidence for selection of alleles conferring resistance to infectious agents. Another point I'd like to make is that the T cell-dependent pathway for IL12 production was found to be deficient in about 25% of people of European descent in studies reported by Jan Mueller-Berghaus from Germany (Mueller-Berghaus et al 2004). He examined the effects of the *IL12B* promoter polymorphism on IL-12 production by DCs after CD40L stimulation. The message from these studies is that many control subjects will not respond in this particular assay.

Casanova: I suppose it depends on the assay, and perhaps on the population tested. In our experimental conditions, the T cell dependent production of IL12 is intact in all healthy controls tested (Filipe-Santos et al 2006).

Morahan: It's like the assay you described earlier on in your paper: co-culture with CD40-ligand transfected cell line (J558).

Casanova: We have tested numerous controls. In the assays I have shown there is variation among individuals, but all controls do respond. The difference between our data may reflect cell type dependent, stimulus dependent, or readout dependent variations in experimental conditions.

Morahan: We also believe that responses of particular *IL12B* genotypes vary according to cell type and stimulus.

Young: What is so important about T cell-dependent IL12? Is it a timing issue?

Casanova: I really don't know. It is surprising to me. Perhaps it suggests that within granulomas there is a need for enhanced production of IL12 in addition to the mycobacterial trigger.

Britton: On that point, Carl Feng and Alan Sher have shown that continued production of IL12 is needed for control of TB infection in the lungs. Recently we have shown that in IL12 knockout mice we can complement the induction of immunity with exogenous IL12, but cannot control continuing infection in the lung in the absence of IL12. It is needed not only for the induction of immunity, but also for the effector arm of immunity.

Hume: I have a comment about genetics we have been discussing, with susceptibility and variable penetrance. It is worth bearing in mind that mutations in Hox genes alter the likelihood of a particular developmental event. The penetrance is essentially plus or minus. This is on an inbred background, so we shouldn't assume that variable penetrance means that there are other genes that modify the penetrance. There is pure chance with genetic events that are stochastic, or responses to environment that we don't really understand. I am kind of interested by the mouse–human dichotomy here. I wonder whether this relates to the fact that human macrophages don't make much nitric oxide (NO), and it is not a major defence. There are all these intracellular pathogens for which we think IFNγ is the major pathway in mice. You are saying that in humans this isn't true; a large number of pathogens are perfectly defendable without this pathway at all. This begs the question, what is the pathway in humans for defence against intracellular pathogens, and do mice have it? Do we have a mouse model that is useful for studying the defence against intracellular pathogens? Should we be studying the IFNγ knockout mouse as a model for human?

Casanova: There are patients with severe infectious diseases caused by many other microorganisms, and they haven't been investigated at the molecular genetic level. Perhaps these investigations will lead to the discovery of other pathways. Regarding inducible nitric oxide synthase (iNOS), I think it is expressed in humans too and we hope to identify human patients with iNOS deficiency some day.

Hume: The gene is expressed at two or three orders of magnitude less, and humans don't make the cofactor, so while the gene is expressed they don't make NO.

Casanova: I thought that one could detect high levels of the protein in tissue sections from TB patients. If induced by mycobacteria, given its role in the mouse, it is likely to play a role in humans too.

Hume: But they don't make tetrahydrobiopterin so they can't make NO. Whatever the function of iNOS is in humans, quantitatively they make vastly less NO than mice do.

Casanova: This is *in vitro* or *ex vivo*. I doubt there are data proving they don't make NO *in situ in vivo*. Again, I think human genetics will provide an answer to this question. If disease-causing mutations were found in the NO production pathway, this would provide conclusive evidence.

Hume: It is pretty hard to make it if you can't make the cofactor.

Young: You could import the cofactor from other cells.

Hume: I have never seen clear evidence that human macrophages make anything like as much NO as mice do. In all the mouse intracellular infections, the IFNγ defence is blocked by NO scavengers. It is utterly NO dependent. The question is, what do humans do if they can't use NO? Presumably, they don't even make NO in the knockout, and in the individuals who don't have the IL12/IFNγ pathway there isn't any NOS, so the question of the cofactor is irrelevant. So how do humans kill intracellular pathogens? It seems that the mouse is not a useful model from what you are saying.

Gros: The mouse is a good model in that mutations in specific pathways cause the same phenotypes as homologous mutations do in humans. Although the ultimate effector mechanism may be different, the regulatory networks are likely to be preserved. In addition, macrophages may have many ways to kill TB. The IFNγ knockout mouse dies within three weeks of TB infection. Although such a rapid death is probably linked to the experimental model used (which in this case does not mirror human TB in the field), the very deleterious effect of absence of IFNγ on host response to TB is clearly validated in both mouse and humans.

Hume: It is a model for a specific subset of pathogens.

Casanova: The mouse model is more sensitive. The immunological phenotypes of human and mouse mutants almost always match, but the infectious phenotypes of mouse mutants are almost always broader. In humans we can identify the genes that are non redundant for protective immunity in a natural ecosystem, whereas in the mouse, because of the experimental conditions, you can push the system and detect more phenotypes that would probably go unnoticed if the mutant animals were in the wild (and were themselves carrying a wild-type background). But there is no contradiction between the two models. The surprise was not that the IL12 IFNγ-deficient patients were vulnerable to mycobacteria: this was predicted by the mouse model. The surprise remains that these patients are resistant to most other microorganisms.

Hume: The question I am finding interesting is how they are resistant to the other organisms. For example, if you treat human monocytes with IFNγ, do you increase killing of *Listeria*?

Casanova: Even if you do, it does not mean it operates *in vivo*. Again, the question of immunity to *Listeria* can also be approached by germline genetics: what genetic defects account for the few cases of human listeriosis?

Goodnow: The real challenge here is emphasized by your missense mutations: unless you know a lot about the gene when you find a missense mutation, if it is only in one kindred that there is an affected individual it is a real challenge. The place where this was made most clear to me was at the Sanger centre where they have a worldwide consortium of about 370 different kindreds with presumed X-linked mental retardation, non-syndromic. They have sequenced all the genes on the X chromosome in each of these boys. They find missense mutations all over the place, but many of them are in RIKEN gene encyclopaedia genes for which we know nothing about their function in the brain. They are missense mutations so it is unclear what they mean. If they see a stop codon there is more likelihood. Over the next few years the tools to be able to go into patient cohorts should become available. We can only test so many mice, but 5 billion people is a significant screen. The real challenge will be that we will likely find all these missense variants that we will be able to make a story of based on microarrays and protein–protein interactions, but then the issue will be convincing others that the story you have concocted is real.

Hume: It's tough to do a backcross series on people!

Kaufman: Do you view these mutations in humans as an unfortunate thing to happen in a family and which is maintained because it is recessive, or as something that happens every once in a while by some kind of problem with a polymerase, or whether these could these be under some kind of selection? That is to say, it could be bad for TB, but it could be good for something else?

Casanova: Most of the Mendelian mutations discovered in the field of primary immunodeficiencies have been recessive. It could take quite some time before they would be counter selected by natural selection. The few but increasingly recognized dominant mutations that I did not discuss today can be counter-selected in populations at a higher rate, but there are still *de novo* mutations. There are even a few multiplex kindreds, with patients in multiple generations. In most cases the children have died before bearing their own children. Theoretically what you say is possible, but I don't have examples in mind of infectious disease-predisposing human mutations for which we know a factor that has acted as a positive selection factor. On the other hand, we know deleterious mutations that have been selected because they protect from some infectious diseases, such as the sickle cell trait and malaria.

Maizels: Where I would expect to see balancing selection is perhaps with alleles like those that Grant Morahan was talking about: quantitative calibrating variants of the same loci, but ones which are extant in polymorphic balance in human populations. When do you find these extreme phenotypes in children?

Casanova: These are not extreme phenotypes. They are the common phenotypes.

Maizels: They are extreme in the sense that they are loss of function.

Casanova: These are not loss of function mutations. They are hypomorphic and affect only one pathway. For most pathways, the present mutations in NEMO have no deleterious impact, whereas loss-of-function mutations in NEMO, as previously shown, are lethal *in utero* precisely because they affect multiple pathways, some of which are critical for fetal development.

Maizels: They contrast with alleles that have very small quantitative changes in the level of particular immunological components. I am interested in whether the same loci in human populations may have in balance alleles which influence susceptibility to a disease by small degrees, conferring an advantage against another infectious disease. There is an interesting example in which people have looked at sheep which have been feral for several hundred years on a Scottish island. These sheep still have polymorphisms in IFNγ SNPs which are associated with protection or susceptibility to infections. Even in a small population over a long time polymorphisms have been maintained (Coltman et al 2000). Should the findings from paediatric setting be used to define candidate genes for much wider analysis taking advantage of the numbers that are out there in the natural experiment of exposure?

Casanova: Possibly. But it is plausible, in my eyes, that the set of genes that control primary infection by TB are not the same set that control reactivation. These sets may not even overlap. We have heard of IL12 variations accounting for adult TB. I look forward to reading the data because this suggests that the same gene would control primary and secondary TB. This is possible. One of the difficulties with the genetics of TB has been that in the past the candidate genes were defined largely based on the mouse model, which is primarily a model of primary infection. Adult TB has little to do with primary infection. My prediction would be that these are different sets of genes.

Karupiah: Why are some of these people also susceptible to salmonella?

Casanova: I don't know. Interestingly, not all these patients with these deficiencies have salmonellosis, but mostly (although not exclusively) the IL12 and IL12R-deficient patients, whereas the IFNγR patients do not have salmonelloses, which suggests that there is an IL12-dependent, IFNγ-independent pathway controlling salmonella. Could this be IL23?

Gros: I have data implicating another gene in this IL12/IFNγ loop, ICSBP. This has exactly that phenotype: mycobacterial susceptibility and salmonella susceptibility (Turcotte et al 2005).

References

Coltman DW, Wilson K, Pilkington JG, Stear MJ, Pemberton JM 2000 A microsatellite polymorphism in the gamma interferon gene is associated with resistance to gastro-intestinal nematodes in a naturally-parasitised population of Soay sheep. Parasitology 122:571–582

Filipe-Santos O, Bustamante J, Haverkamp MH et al 2006 X-linked susceptibility to mycobacteria is caused by mutations in NEMO impairing CD40-dependent IL-12 production. J Exp Med 203:1745–1759

Muller-Berghaus J, Kern K, Paschen A et al 2004 Deficient IL-12p70 secretion by dendritic cells based on IL12B promoter genotype. Genes Immun 5:431–434

Turcotte K, Gauthier S, Tuite A, Mullick A, Malo D, Gros P 2005 A mutation in the Icsbp/IRF-8 gene causes susceptibility to infections and a CML-like syndrome in BXH-2 mice. J Exp Med 201:881–890

The genetic control of susceptibility to *Mycobacterium tuberculosis*

W. J. Britton*†, S. L. Fernando*, B. M. Saunders*†, R. Sluyter‡ and J. S. Wiley‡

** Centenary Institute of Cancer Medicine & Cell Biology, Locked Bag 6, Newtown, 2042, † Department of Medicine, Central Clinical School and ‡ Nepean Hospital, Western Central School, University of Sydney, 2006, NSW, Australia*

Abstract. *Mycobacterium tuberculosis* is one of the most successful human pathogens, surviving in latent foci of infection in one third of humanity, yet causing lung necrosis in sufficient individuals to ensure its transmission. Each stage of the host response to *M. tuberculosis* is under genetic control, including the initial encounter with mycobacteria by macrophages, epithelial cells and dendritic cells in the lung, induction of the inductive T cell response, and killing by activated macrophages within granulomas. Although environmental factors are important determinants of progression to disease, there is a genetic component underlying susceptibility to tuberculosis (TB), the basis of which may vary in different populations. Recent studies using a variety of methods have defined a number of susceptibility alleles for the development of active TB. Many of these influence macrophage responses to mycobacteria. We have studied the influence of loss of function polymorphisms in the human *P2X7* gene on the capacity of macrophages to kill *M. tuberculosis.* Activation of the $P2X_7$ receptor, an ATP-gated Ca^{2+} channel, leads to the formation of pores, the activation of phospholipase D, and the induction of apoptosis with death of the infecting mycobacteria. Macrophages from subjects who are heterozygote, homozygote or compound heterozygote for these polymorphisms fail to undergo apoptosis and show partial or complete inhibition of mycobacterial killing. One of these non-functioning polymorphisms was significantly associated with increased susceptibility to TB disease, particularly extrapulmonary disease, in two unrelated cohorts of TB patients. Insights into the genetic regulation of susceptibility to human TB may identify novel methods for controlling latent *M. tuberculosis* and reducing the burden of tuberculosis.

2007 Decoding the genomic control of immune reactions. Wiley, Chichester (Novartis Foundation Symposium 281) p 79–92

Mycobacterium tuberculosis is one of the most successful pathogens of humans. Approximately one third of the world's population harbours latent infection with *M. tuberculosis* with no ill effect. Reactivation of infection in a small proportion of these subjects leads to progressive lung inflammation with necrosis, and the resulting cavitation permits aerosol spread of the mycobacterium to others. Following acute infection with *M. tuberculosis*, 5% of subjects will develop primary tuberculosis disease (TB); however, the majority contain the infection within granulomatous lesions in the lung, but retain the life-long risk for reactivation of the dormant infection. This cycle currently results in nine million new cases of TB, with two million deaths per annum (Dye 2006). Between 2000 and 2020, it is estimated that one billion humans will be newly infected with *M. tuberculosis*, and there will be 200 million active cases of TB with 35 million deaths. The bulk of this active TB disease is caused by reactivation of dormant infection and a recent longitudinal study in Liverpool, NSW, showed that the lifetime risk of reactivation for a 25 year old with latent TB infection (LTBI) is 7.3% (Marks et al 2000). Despite the availability of effective antimicrobial therapy, TB remains poorly controlled in the majority of endemic countries, and ranks second only to Human Immunodeficiency Virus (HIV) as a cause of death from an infective organism worldwide (Corbett et al 2003).

There are multiple determinants of the outcome of *M. tuberculosis* infection, including factors affecting the host, the mycobacterium and the social environment. Recent studies have demonstrated variation in the virulence of clinical isolates of *M. tuberculosis* (Reed et al 2004) and multidrug resistant strains of *M. tuberculosis* have increased in prevalence worldwide and may be responsible for treatment failure. Poverty, overcrowding, war and social disorder may limit or disrupt the treatment services. Factors influencing the host immune response are also critical in determining the response to *M. tuberculosis* infection and the risk of reactivation. These include concurrent infections, particularly HIV, which is associated with rapid progression of primary TB infection and reactivation of LTBI in 50% of co-infected subjects (Corbett et al 2003). Other infections, including measles, in addition to protein/calorie malnutrition and prior Bacille Calmette-Guérin (BCG) immunization also influence the effectiveness of the host response. In addition, genetic factors play a significant role in determining the outcome of infection.

Immune responses to *M. tuberculosis*

Genetic factors may influence each stage of the host response to *M. tuberculosis* infection (Flynn & Chan 2001). The initial encounter with mycobacteria occurs in the peripheral lung where alveolar macrophages and dendritic cells take up the organisms. This phagocytosis occurs through multiple cell surface molecules, including complement receptors, mannose receptors, DC-SIGN and other C-type lectins. Toll-like receptors (TLR) on the surface of macrophages and dendritic cells are activated by mycobacterial products. This stimulates the production of pro-inflammatory cytokines and the phenotypic maturation and migration of mycobacteria-infected DCs to the draining lymph nodes. The immunological control of *M. tuberculosis* is dependant on the activation of antigen-specific $CD4^+$ and $CD8^+$ T cells in the draining lymph nodes. The differentiation of $CD4^+$ T cells into Th1-like T cells is dependent on MHC class II-restricted presentation of mycobacterial peptides by infected dendritic cells to $CD4^+$ T cells, in the presence of the cytokines interleukin (IL)12, IL18 and IL23. These activated Th1-like $CD4^+$ T cells produce interferon (IFN)γ, which activates phagolysomal fusion in macrophages and the production of antimycobacterial molecules, including reactive oxygen and nitrate intermediates (Flynn & Chan 2001). Other $CD4^+$ T cell-derived cytokines, including lymphotoxin α and TNF, are also important for the activation of macrophages and the formation and maintenance of granulomas, which contain the infection to the lung (Roach et al 2001). *M. tuberculosis* infected dendritic cells also activate $CD8^+$ T cells through the MHC class I pathway. These $CD8^+$ T cells are also recruited to the sites of infection and contribute to the mycobacterial killing.

Mycobacteria have developed multiple mechanisms to evade killing by macrophages, so that activation of human macrophages by IFNγ fails to eradicate the *M. tuberculosis* infection (Flynn & Chan 2001). Further, transcriptional responses of *M. tuberculosis* to the intracellular environment result in the organism entering a dormant state of infection. Control of this persisting infection requires the maintenance of sustained memory T cell response to mycobacterial antigens, and the preservation of granulomas around the infected macrophages. Studies of the genetic control of mycobacterial infections in animals have focused on the acute response to infection. By contrast, the majority of human studies have examined the genetic factors influencing reactivation of LTBI.

Decoding the genetic control of mycobacterial infection

Evidence for the genetic influence on mycobacterial infection include the increased concordance for tuberculosis and leprosy between monozygotic twins as compared to dizygotic twins, and the different susceptibility of ethnic groups to tuberculosis, independent of social and environmental factors (reviewed in Bellamy 2003). In addition, rare mutations in five genes controlling the IL12–IFNγ axis have been found to confer increased susceptibility to progressive infection with either the vaccine strain, *Mycobacterium bovis*, (BCG) or environmental mycobacteria, *Mycobacterium avium* (reviewed in Casanova & Abel 2002). This highlights the importance of the IFNγ signalling pathway for effective mycobacterial immunity; however, these do not explain genetic susceptibility to TB infection in the wider population. The latter is probably due to the polygenic effect of polymorphisms affecting multiple genes acting at different stages of the host response to intracellular bacterial infection. A variety of approaches have been used to define these host susceptibility genes, including case control studies analysing defined polymorphisms in candidate genes and family based genome-wide linkage studies (Bellamy 2003).

TABLE 1 Genes linked with increased susceptibility to or protection against tuberculosis disease identified by case control studies

Gene	*Function*
Associations in multiple studies	
HLA-DR2, DQB, class I	Antigen presentation
SCL11A1 (nRAMP-1)	Macrophage function
VDR (vitamin D receptor)	Macrophage activation
MBL (mannose binding ligand)	Innate immunity
Associations in single studies	
TLR2 (Toll-like receptor 2)	Innate immunity
SPA, SPD (surfactant protein A and D)	Adhesion
IL1Rα	Inflammation
IL12, IL12RB1 (IL12, receptor)	T cell activation
IFNγ (interferon γ)	Macrophage activation
P2X7 ($P2X_7$ purinergic receptor)	Macrophage activation
SP110 (SP110)	Possible transcription factor

Primary references listed in Fernando & Britton (2006), Bellamy (2003) and Casanova & Abel (2002).

Multiple case-control studies have identified alleles in at least 14 genes which confer increased susceptibility or protection against clinical TB (reviewed in Fernando & Britton 2006). These genes influence the innate immune responses to infection with mycobacteria, the activation of Th1-like $CD4^+$ T cells, or the stimulation of infected macrophages (Table 1). Only a small number of these candidate susceptibility genes have been associated with TB in multiple independent studies and the effect of individual polymorphisms is generally modest. These include the genes encoding NRAMP1 (natural resistance-associated macrophage protein 1, also termed solute carrier family 11A-1 [*SCL11A1*]), the vitamin D receptor (VDR), HLA class I and class II molecules, and mannose binding lectin (MBL). Numerous polymorphisms in *NRAMP1*, which was first found to confer resistance to intracellular pathogens in mice, have been associated with susceptibility to TB in different human populations (Bellamy et al 1998). Moreover, polymorphisms in *NRAMP1* which confer increased susceptibility to TB in one population, such as The Gambia (Bellamy et al 1998), are not associated with susceptibility in other populations, such as Cambodia and Malawi (Delgado et al 2002, Fitness et al 2004). Therefore, both the genes and the variants within individual genes, which confer increased susceptibility on TB, vary among different populations.

Few studies have examined genetic susceptibility to extrapulmonary TB. However, in one study polymorphisms in *MBL* conferred protection against meningeal TB (Hoal-Van Helden et al 1999). Some of the TB susceptibility genes, including *VDR*, *HLA-DR2* and *NRAMP1*, are also associated with the increased risk of leprosy in case control studies, supporting their association with the control of mycobacterial infections (Alcais et al 2005, Fernando & Britton 2006).

Environmental factors may influence the effect of a genetic susceptibility allele in individuals. One study of TB in Gujarats living in London, found that an association with alleles of *VDR* with TB only in the presence of low or undetectable levels of vitamin D (Wilkinson et al 2000). Concurrent HIV infection may also lead to loss of association of a susceptibility gene with TB. Therefore, variation in environmental factors may be responsible for some of the differences observed in susceptibility studies in individual genes in different populations.

Genome-wide screening using family studies is an alternate approach to identify important loci linked to TB. This has the advantage of discovering previously unidentified pathways involved in the pathogenesis of disease. However, large populations are required to fulfil the criteria for genome-wide statistical significance. A study of African subjects with pulmonary TB suggested linkage of TB with two loci on chromosomes 15q L213 and chromosome X (Bellamy et al 2000), although neither reached statistical significance. Interestingly, no linkage was found with the regions encoding the *NRAMP1* and *HLA-DR*, which are associated with TB in African subjects in case control studies. More recently, a locus on chromosome 8q12-q13 was significantly linked to predisposition to pulmonary TB in Moroccan subjects (Baghdadi et al 2006). Similar techniques in mice have identified that the *sst-1* locus on chromosome 1 is linked to progressive pulmonary TB, and the candidate gene, *IPR1*, is up-regulated in macrophages from resistant mice during *M. tuberculosis* infection (Pan et al 2005). Polymorphisms in the human homologue of *IPR1*, *SP110* have recently been associated with increased susceptibility to TB in African subjects (Tosh et al 2006).

Purinergic P2X$_7$ receptors

P2X receptors are a family of seven-transmembrane ion channels, which are activated by extracellular ATP to stimulate the rapid influx of cations (Khakh & North 2006). They are widely distributed, particularly in the nervous system, and are implicated in a range of physiological and pathological responses. The P2X$_7$ receptor is expressed on the surface of all leukocytes, with maximum expression on monocytes, which is further increased by maturation of monocytes to macrophages and their stimulation by IFNγ (Gu et al 2001). Activation of P2X$_7$ causes an immediate opening of a cation-selective channel and the influx of Ca^{2+}, followed by the formation of a pore which permeabilizes the cell allowing the entry of fluorescent dyes, such as ethidium$^+$ (Wiley et al 1998). This permits analysis of P2X$_7$ function by time-resolved flow cytometry. Activation of P2X$_7$ with ATP also results in the induction of the caspase cascade, with resultant apoptosis of the cell (Ferrari et al 1999) and the activation of phospholipase D

(PLD) (el-Moatassim & Dubyak 1992). This is associated with the release of IL1β and IL18, and macrophages from $P2X_7$-deficient mice show deficient release of these cytokines after ATP activation and reduced inflammation in models of arthritis (Labasi et al 2002). $P2X_7$-deficient mice also exhibit reduced periosteal bone formation, implicating $P2X_7$ in normal bone homeostasis (Ke et al 2003). Lammas and colleagues demonstrated that activation of $P2X_7$ on mycobacteria-infected macrophages induced apoptosis, which was accompanied by mycobacterial killing (Lammas et al 1997). This effect is dependent on ATP-mediated activation of PLD (Kusner & Adams 2000) and Ca^{2+}-dependent promotion of phagolysosomal fusion, a process essential for mycobacterial killing (Fairbairn et al 2001).

Genetic variation in $P2X_7$ receptors

We have used time-resolved flow cytometry to identify subjects with loss of function single nucleotide polymorphisms (SNPs) in the $P2X_7$ receptor gene (*P2X7*), and then examined the effects of these SNPs on the capacity of infected macrophages to kill BCG and *M. tuberculosis*. Several SNPs in the coding regions of *P2X7* impair ATP-mediated function. The most common of these, with an allele frequency of 0.17 in Caucasians, is the 1513A → C SNP, which causes an amino acid change, E496A, in the putative ankyrin repeat motif of the C-terminus of the $P2X_7$ receptor (Gu et al 2001). Ankyrin repeats play a major role in anchoring proteins to the membrane or cytoskeleton, as well as in protein folding and protein–protein interaction, and this region may contribute to the polymerization of the receptor. The 1729T → A polymorphism results in amino acid change, I568N, in the C-terminus of the $P2X_7$ gene, and this abolishes trafficking of the receptor to the cell surface (Wiley et al 2003). By contrast, the amino acid change, R307Q, caused by the 946G → A polymorphism, abolishes binding of ATP to the extracellular domain of $P2X_7$ (Gu et al 2004). The 151+1G → T SNP, generates a nonsense transcript which prevents synthesis of $P2X_7$ (Skarratt et al 2005).

Effect of genetic variation in $P2X_7$ on mycobacterial killing

Loss-of-function polymorphisms in *P2X7* are associated with impaired ability to kill mycobacteria via ATP *in vitro* (Saunders et al 2003, Fernando et al 2005). Monocytes from relevant subjects were differentiated into macrophages by adhesion, activated with IFNγ, and then infected with *M. bovis* BCG at an MOI of 5. After two days, the cells were activated by brief exposure to ATP for 20 min and cultured for 16 h. The proportion of macrophages undergoing apoptosis was measured by staining for Annexin-V, and the degree of BCG killing relative to non-

ATP activated macrophages was determined by quantitative microbiology. In macrophages from subjects homozygous for the 1513A → C SNP in $P2X_7$, ATP-induced apoptosis was reduced from 30% in normal subjects to zero, and the inhibition of BCG growth of 1 × log10 observed in normal subjects was fully ablated. Macrophages from 1513A → C heterozygous subjects showed a 50% reduction in ATP-induced BCG killing and apoptosis (Saunders et al 2003). A similar effect was observed in macrophages from subjects who were heterozygous for the other, less frequent, loss of function SNPs, viz. 1729T → A, 946G → A and 151+1G → T. The effect of these loss-of-function polymorphisms on ATP-responsiveness was additive. Macrophages from individuals who were compound heterozygous with different combinations of these four loss-of-function polymorphism had ablated ATP-responsiveness and were unable to kill either BCG or *M. tuberculosis* (Fernando et al 2005).

Influence of genetic variation in $P2X_7$ on susceptibility to tuberculosis

To determine whether polymorphisms in *P2X7* confer susceptibility to TB, we conducted two independent case-control studies in subjects predominately from South-East Asia. The frequency of the four loss-of-function SNPs was determined in normal subjects from this region, and only the 1513A → C SNP, with an allele frequency of 0.24, was present in > 1% of South East Asians. Therefore, we examined the association of this SNP with the reactivation of LTBI in a nested case-control study of the Liverpool cohort of refugees in NSW (S. L. Fernando, personal communication). This cohort included 8609 subjects with LTBI confirmed by positive tuberculin skin test and normal chest X-ray at initial post-arrival screening (Marks et al 2000). Cross analysis with the state TB registry indicated that 119 of the subjects with LTBI had developed active TB over a median surveillance period of 13 years. DNA samples were obtained from 86 of the 90 TB patients contactable, and from 167 matched controls from the same cohort. Univariate analysis of genotyping data for the 1513A → C SNP revealed a significant association between the 1513C allele and TB in this cohort with an OR of 1.7 (95% CI 1.0–2.9, $P < 0.05$). This association was evident in the patients with extrapulmonary TB (30/86) with an OR of 4.0, (95% CI 1.7–9.3, $P \leq 0.001$), but was not significant in pulmonary disease. Logistic regression analysis demonstrated that the strong association between the 1513C allele and extrapulmonary TB was not affected by the age, gender, BCG status or the degree of tuberculin positivity of the subjects.

The association between the 1513A → C polymorphism and active TB was then tested in a second independent cohort of 99 recently diagnosed, HIV-negative TB patients (50 pulmonary and 49 extrapulmonary) and control subjects with LTBI ascertained in Sydney. In this cohort, univariate analysis revealed a significant association between the 1513C allele and active TB (OR 1.9, 95% CI 1.1–3.4, $P < 0.05$). As in the Liverpool cohort, the association was only evident for extra-pulmonary disease with an OR of 3.0 (95% CI 1.5–6.0, $P < 0.01$), and this strong association was independent of age and gender.

The functional relationship between the 1513C allele and the capacity to kill mycobacteria was examined in TB patients and controls in the second cohort. Monoctye-derived macrophages from TB patients and control subjects, who were heterozygous for the 1513A → C polymorphism in *P2X7*, displayed a significant impairment in their ability to kill BCG following ATP-stimulation, as compared to *P2X7* WT macrophages ($P < 0.0001$). The ATP-mediated mycobacterial killing was ablated in macrophages from 1513C homozygous subjects. This reduction in killing was associated with a significant reduction in ATP-induced apoptosis in macrophages from heterozygous and homozygous TB patients and control subjects. There was a strong correlation between the levels of ATP-induced apoptosis and BCG killing in macrophages from TB patients and control subjects ($r_s^2 = 0.65$, $P < 0.0001$).

These functional studies highlight two important features of genetic control of TB. First, genetic susceptibility to reactivation of LTBI as a complex trait involves the interactions of multiple genes, so that heterozygosity or homozygosity for the 1513C allele alone was not sufficient for reactivation of TB, presumably because of the influence of other genes. Second, other genetic variations are likely to contribute to the capacity for ATP-mediated killing of mycobacteria. Although the 1513C allele occurred in over 80% of subjects with a major reduction in mycobacterial killing, we identified subjects, who were wild-type at 1513A, but had markedly reduced killing capacity. We sequenced the 13 exons of *P2X7* in 12 such subjects, but failed to detect other, known loss-of-function polymorphisms in *P2X7*. Different polymorphisms in *P2X7* were identified in some subjects and we are evaluating their effect on $P2X_7$ function, however the failure in mycobacterial killing could not be attributed to genetic variation in *P2X7* in more than half of these subjects. Therefore, polymorphisms in genes encoding other components of the $P2X_7$–PLD signalling pathway may influence macrophage killing of mycobacteria and susceptibility to TB.

Discussion

This study strengthens the role of the $P2X_7$ receptor and ATP-mediated apoptosis of infected macrophages in the control of *M. tuberculosis* infection. A separate case-control study in The Gambia identified the significant association of a protective effect against pulmonary TB with a SNP at position −792 in the putative promoter region of *P2X7* (Li et al 2002). Interestingly, no association was found between susceptibility to TB and the 1513A → C SNP in this population, but this may be because the Gambian TB patients were restricted to pulmonary TB, and this form of TB was not associated with the 1513C allele in Asians. Moreover, the frequency of the 1513C allele was low (0.08) in Gambian subjects, compared to South-East Asians (0.24).

These studies demonstrate characteristics common to other gene association studies with TB. First, the pattern of association with TB for an individual gene or allele may vary between populations. For example, polymorphisms in *NRAMP1* conferred increased risk of pulmonary TB in Gambians (Bellamy et al 1998), a protective effect against TB in Cambodians (Delgado et al 2002), and had no influence on the susceptibility to TB in Brazilians (Shaw et al 1997). In addition, polymorphisms in the gene for MBL were associated with protection against TB in South African (Hoal-Van Helden et al 1999) populations, but increased susceptibility to TB in South Indians (Selvaraj et al 1999). These varying results in diverse racial groups may reflect not only differences in allele frequency, but linkage disequilibrium with other, as yet unidentified, susceptibility loci in these different populations. Second, there may be differences in the genetic variants conferring susceptibility to pulmonary and extrapulmonary forms of the disease. For example, TB meningitis, a major form of extrapulmonary TB in children, was associated with a polymorphism in *MBL* in South Africa children (Hoal-Van Helden et al 1999) and variants of Toll-like receptors in Vietnamese patients (Hawn, personal communication 2006). This suggests that the genes identified by these associations are critical to the immune responses limiting the spread of the initial focus of *M. tuberculosis* infection from the lung to extrapulmonary sites.

In summary, the association of genetic variants in the $P2X_7$ receptor with TB in different populations highlights the importance of the $P2X_7$-PLD pathway in the macrophage-mediated control of *M. tuberculosis* infection, and suggests that activating this pathway would be a novel approach to enhancing macrophage clearance of infection and reducing the risk of subsequent TB in subjects with LTBI.

Acknowledgements

We thank Dr G. Marks and his staff for their expert involvement in these studies, which were supported by the National Health and Medical Research Council of Australia, the Australian Respiratory Council, and the New South Wales Department of Health through its research infrastructure grant to the Centenary Institute.

References

Alcais A, Mira M, Casanova JL, Schurr E, Abel L 2005 Genetic dissection of immunity in leprosy. Curr Opin Immunol 17:44–48

Baghdadi JE, Orlova M, Alter A et al 2006 An autosomal dominant major gene confers predisposition to pulmonary tuberculosis in adults. J Exp Med 203:1679–1684

Bellamy R 2003 Susceptibility to mycobacterial infections: the importance of host genetics. Genes Immun 4:4–11
Bellamy R, Ruwende C, Corrah T et al 1998 Variations in the NRAMP1 gene and susceptibility to tuberculosis in West Africans. N Engl J Med 338:640–604
Bellamy R, Beyers N, McAdam KP et al 2000 Genetic susceptibility to tuberculosis in Africans: a genome-wide scan. Proc Natl Acad Sci USA 97:8005–8009
Casanova JL, Abel L 2002 Genetic dissection of immunity to mycobacteria: the human model. Annu Rev Immunol 20:581–620
Corbett EL, Watt CJ, Walker N et al 2003 The growing burden of tuberculosis: global trends and interactions with the HIV epidemic. Arch Intern Med 163:1009–1021
Delgado JC, Baena A, Thim S, Goldfeld AE 2002 Ethnic-specific genetic associations with pulmonary tuberculosis. J Infect Dis 186:1463–1468
Dye C 2006 Global epidemiology of tuberculosis. Lancet 367:938–940
el-Moatassim C, Dubyak GR 1992 A novel pathway for the activation of phospholipase D by P2z purinergic receptors in BAC1.2F5 macrophages. J Biol Chem 267:23664–23673
Fairbairn IP, Stober CB, Kumararatne DS, Lammas DA 2001 ATP-mediated killing of intracellular mycobacteria by macrophages is a P2X(7)-dependent process inducing bacterial death by phagosome-lysosome fusion. J Immunol 167:3300–3307
Fernando SL, Britton WJ 2006 Genetic susceptibility to mycobacterial disease in humans. Immunol Cell Biol 84:125–137
Fernando SL, Saunders BM, Sluyter R et al 2005 Gene dosage determines the negative effects of polymorphic alleles of the P2X7 receptor on adenosine triphosphate-mediated killing of mycobacteria by human macrophages. J Infect Dis 192:149–155
Ferrari D, Los M, Bauer MK et al 1999 P2Z purinoreceptor ligation induces activation of caspases with distinct roles in apoptotic and necrotic alterations of cell death. FEBS Lett 447:71–75
Fitness J, Floyd S, Warndorff DK et al 2004 Large-scale candidate gene study of tuberculosis susceptibility in the Karonga district of northern Malawi. Am J Trop Med Hyg 71:341–349
Flynn JL, Chan J 2001 Immunology of tuberculosis. Annu Rev Immunol 19:93–129
Gu BJ, Zhang W, Worthington RA et al 2001 A Glu-496 to Ala polymorphism leads to loss of function of the human P2X7 receptor. J Biol Chem 276:11135–11142
Gu BJ, Sluyter R, Skarratt KK et al 2004 An Arg307 to Gln polymorphism within the ATP-binding site causes loss of function of the human P2X7 receptor. J Biol Chem 279:31287–31295
Hoal-Van Helden EG, Epstein J, Victor TC et al 1999 Mannose-binding protein B allele confers protection against tuberculous meningitis. Pediatr Res 45:459–464
Ke HZ, Qi H, Weidema AF et al 2003 Deletion of the P2X7 nucleotide receptor reveals its regulatory roles in bone formation and resorption. Mol Endocrinol 17:1356–367
Khakh BS, North RA 2006 P2X receptors as cell-surface ATP sensors in health and disease. Nature 442:527–532
Kusner DJ, Adams J 2000 ATP-induced killing of virulent Mycobacterium tuberculosis within human macrophages requires phospholipase D. J Immunol 164:379–388
Labasi JM, Petrushova N, Donovan C et al 2002 Absence of the P2X7 receptor alters leukocyte function and attenuates an inflammatory response. J Immunol 168:6436–6445
Lammas DA, Stober C, Harvey CJ et al 1997 ATP-induced killing of mycobacteria by human macrophages is mediated by purinergic P2Z(P2X7) receptors. Immunity 7:433–444
Li CM, Campbell SJ, Kumararatne DS et al 2002 Association of a polymorphism in the P2X7 gene with tuberculosis in a Gambian population. J Infect Dis 186:1458–1462
Marks GB, Bai J, Simpson SE, Sullivan EA, Stewart GJ 2000 Incidence of tuberculosis among a cohort of tuberculin-positive refugees in Australia: reappraising the estimates of risk. Am J Respir Crit Care Med 162:1851–1854

Pan H, Yan BS, Rojas M et al 2005 Ipr1 gene mediates innate immunity to tuberculosis. Nature 434:767–772
Reed MB, Domenech P, Manca C et al 2004 A glycolipid of hypervirulent tuberculosis strains that inhibits the innate immune response. Nature 431:84–87
Roach DR, Briscoe H, Saunders B et al 2001 Secreted lymphotoxin-alpha is essential for the control of an intracellular bacterial infection. J Exp Med 193:239–246
Saunders BM, Fernando SL, Sluyter R, Britton WJ, Wiley JS 2003 A loss-of-function polymorphism in the human P2X7 receptor abolishes ATP-mediated killing of mycobacteria. J Immunol 171:5442–5446
Selvaraj P, Narayanan PR, Reetha AM 1999 Association of functional mutant homozygotes of the mannose binding protein gene with susceptibility to pulmonary tuberculosis in India. Tuber Lung Dis 79:221–227
Shaw MA, Collins A, Peacock CS et al 1997 Evidence that genetic susceptibility to Mycobacterium tuberculosis in a Brazilian population is under oligogenic control: linkage study of the candidate genes NRAMP1 and TNFA. Tuber Lung Dis 78:35–45
Skarratt KK, Fuller SJ, Sluyter R et al 2005 A 5′ intronic splice site polymorphism leads to a null allele of the P2X7 gene in 1–2% of the Caucasian population. FEBS Lett 579:2675–2678
Tosh K, Campbell SJ, Fielding K et al 2006 Variants in the SP110 gene are associated with genetic susceptibility to tuberculosis in West Africa. Proc Natl Acad Sci USA 103:10364–10368
Wiley JS, Gargett CE, Zhang W, Snook MB, Jamieson GP 1998 Partial agonists and antagonists reveal a second permeability state of human lymphocyte P2Z/P2X7 channel. Am J Physiol 275:C1224–1231
Wiley JS, Dao-Ung LP, Li C et al 2003 An Ile-568 to Asn polymorphism prevents normal trafficking and function of the human P2X7 receptor. J Biol Chem 278:17108–17113
Wilkinson RJ, Llewelyn M, Toossi Z et al 2000 Influence of vitamin D deficiency and vitamin D receptor polymorphisms on tuberculosis among Gujarati Asians in west London: a case-control study. Lancet 355:618–621

DISCUSSION

Casanova: Are the two populations that you have studied related ethnically?

Britton: Partially. The first which we followed longitudinally, was 90% Vietnamese and Cambodian. The second included TB patients recruited over a three year period in Sydney, and of these 60% were from Southeast Asia, including Vietnam and Cambodia, China and the Philippines.

Gros: Is the effect of the receptor polymorphisms on killing of mycobacteria by macrophages dependent on IFNγ?

Britton: Yes. IFNγ leads to up-regulation of the $P2X_7$ receptor, but this effect varies among individuals. Maturation of macrophages from monocytes is sufficient for partial up-regulation of the receptor, but exposure to IFNγ leads to maximum expression. In subsequent studies we have looked at the effects of different combinations of cytokines on the expression of the receptor. In some individuals, macrophage differentiation is sufficient to cause maximal expression, but in others, a combination of IFNγ and TNF is required.

Gros: Do you see an effect of IFNγ on the activity of this protein?

Britton: We think the effect of IFNγ is on the level of expression of the receptor. We measure the level of expression with a monoclonal antibody, as well as measuring the function through ethidium uptake.

Casanova: These were case control studies. How could you match your controls if you didn't know the ethnic group of your patients? This probably introduces some biases.

Britton: In the nested case control study of subjects with LTBI, we matched TB patients with controls from the same cohort, which was 90% homogeneous ethnically, according to their degree of tuberculin reactivity. For the second study, we recruited control subjects from TB clinics who had been exposed to tuberculosis and were tuberculin positive, About 60% of these were similar ethnically.

Morahan: The frequency of the defective allele is relatively high in the Vietnamese groups. Do you know why?

Britton: $P2X_7$ has a pro-inflammatory properties and the defective allele may reduce this effect. Drug companies have developed inhibitors against $P2X_7$, because activation of this receptor leads to increased production of IL1 and IL18.

Hume: The screen is set up so that it is ATP dependent. What other molecules will provide the triggering stimulus for the killing process? Presumably, those pathways downstream of IFNγ would be the place where you would look for susceptibility loci.

Britton: The tumour necrosis factor (TNF) pathway has been examined, but reduced TNF production or polymorphisms in the TNF gene have not been associated with TB. There is debate about the TNF promoter polymorphisms, but they haven't been particularly informative in either TB or leprosy. Activated vitamin D is produced by macrophages and activates macrophages to kill mycobacteria, and variants in the vitamin D receptor have been associated with susceptibility to TB and leprosy.

Cyster: Where is the ATP coming from?

Britton: This is the major question concerning $P2X_7$ function. The ATP concentrations used to activate $P2X_7$ are higher than the concentrations required for the P2Y receptors. But this is the same for all of the effects of $P2X_7$ receptor, including its proven requirement for bone homeostasis and inflammatory responses.

Cyster: Is there a model as to how this works?

Britton: We are examining the effects of $P2X_7$ in T cell/macrophage co-cultures using macrophages from homozygous and heterozygous individuals, to determine whether there is reduced mycobacterial killing following activation of defective $P2X_7$.

Goodnow: In terms of the high frequency of this one particular allele, do you have any idea about its function?

Britton: We don't have any definitive answer. In the heterozygote subjects there is reduced IL1 and IL18 production, which could be modulating the inflammatory response.

Turner: Is there any relationship between the ability to kill off the mycobacterium and induction of an adaptive immune response, in terms of the magnitude? If you kill the macrophages and TB releases antigen, does this result in a bigger response versus sequestering the antigen?

Britton: We haven't done these studies. Stephan Kaufmann's group (Winau et al 2006) has shown that apoptosis may be releasing vesicles of the membrane of infected cells and these vesicles contain mycobacterial fragments. Purified apoptotic vesicles can prime DCs for CD8 T cell responses and immunise for protection against TB. It is not clear whether or not this happens during infection *in vivo*. There is no doubt that inducing death through apoptosis is the way to control TB in human macrophages.

Turner: I was wondering about the implications for the establishment of memory. If you aren't getting a good immune response going on in the first place you are not killing off the bugs. Thus CD4 memory might wane over time.

Britton: It's not easily testable. We have acquired $P2X_7$ knockout mice, but these do not show increased susceptibility to TB. But the $P2X_7$ knockout is made on a BL/6 background, and there is a polymorphism in BL/6 mice that reduces $P2X_7$ function compared to other strains including BALB/c. This may explain why the $P2X_7$ knockout doesn't show increased susceptibility to TB. This also suggests there may be differences in macrophage killing mechanisms between mouse and human.

Hume: Do you see differences in the inducible cytokine profiles in response to mycobacteria, dependent on that mutation and irrespective of ATP?

Britton: We have examined class II expression and TNF production after ATP stimulation and shown that these are unaffected by $P2X_7$ function.

Goodnow: The general message I get from putting Jean-Laurent Casanova's paper together with yours is that when you find these missense amino acid substitutions, unless you have a good functional bioassay it will be tough to interpret it much.

Britton: We have identified a number of loss-of-function polymorphisms in this gene, but these alleles must be of sufficient frequency for use in association studies. It has been helpful to have an *in vitro* model to help select which polymorphisms to study at a population level.

Casanova: You speak of large population samples, but in your two case control studies the extrapulmonary cases were 30 and 50, which are relatively small samples.

Britton: Perhaps if we had a larger group of TB patients, we might have found an association with pulmonary TB as well as extrapulmonary TB.

Hume: Could you safely administer a $P2X_7$ agonist?

Britton: There are some variants of ATP which have higher affinity for $P2X_7$ which could be considered. The effect of the agonist would be maximal at sites of IFNγ expression, such as in granulomas containing mycobacteria, and this would focus the effect of the $P2X_7$ agonist. We have also considered delivery into the lung to shorten the length of chemoprophylaxis with isoniazid. The pharmaceutical industry has developed $P2X_7$ inhibitors for their anti-inflammatory effects, and there is potential to develop agonists.

Cook: What is the mortality of extrapulmonary TB?

Britton: The mortality is higher than for pulmonary TB in both in children and adults.

Cook: I am wondering if untreated, is it high enough to affect the conclusions from this study, since those carrying mutations would be under-represented in the analysed component of the cohort?

Britton: We don't know whether this has affected their selection.

Reference

Winau F, Weber S, Sad S et al 2006 Apoptotic vesicles crossprime CD8 T cells and protect against tuberculosis. Immunity 24:105–117

Th2 lymphoproliferative disorders resulting from defective LAT signalosomes

Bernard Malissen, Ying Wang, Michael Mingueneau and Marie Malissen

Centre d'Immunologie de Marseille-Luminy, Université de la Méditerranée, Case 906, 13288 Marseille Cedex 9, INSERM, U631, 13288 Marseille Cedex 9 and CNRS, UMR6102, 13288 Marseille Cedex 9, France

Abstract. LAT (linker for activation of T cells) is an integral membrane adaptor protein that constitutes in T cells a major substrate of the ZAP-70 protein tyrosine kinase. LAT coordinates the assembly of a multiprotein signalling complex through phosphotyrosine-based motifs present within its intracytoplasmic segment. The resulting 'LAT signalosome' links the TCR to the main intracellular signalling pathways that regulate T cell development and T cell function. Early studies using transformed T cell lines suggested that LAT acts primarily as a positive regulator of T cell receptor (TCR) signalling. The partial or complete inhibition of T cell development observed in several mouse lines harbouring mutant forms of LAT was congruent with that view. More recently, LAT 'knock-ins' harbouring point mutations in the four COOH-terminal tyrosine residues, were found to develop lymphoproliferative disorders involving polyclonal T cells that produced high amounts of T helper-type 2 (Th2) cytokines. This unexpected finding revealed that LAT also constitutes a negative regulator of TCR signalling and T cell homeostasis. As discussed, the available data underscore that a novel immunopathology proper to defective LAT signalosome is likely taking shape.

2007 Decoding the genomic control of immune reactions. Wiley, Chichester (Novartis Foundation Symposium 281) p 93–102

LAT (linker for activation of T cells) is an integral membrane adaptor protein that constitutes in T cells a major substrate of the ZAP-70 protein tyrosine kinase (PTK) (Zhang et al 1998). Adaptor proteins lack both enzymatic and transcriptional activities and act as molecular scaffolds through which multiprotein signalling complexes are transiently assembled. LAT possesses a juxtamembrane CXXC palmitolylation motif (where C denotes cysteine and X any amino acid). Palmitoylation stabilizes the association of LAT with the plasma membrane and targets it to lipid rafts. The essential role LAT plays in T cell signalling was first deduced from the analysis of LAT-deficient variants of the Jurkat T cell line (Finco et al

1998, Zhang et al 1999). Subsequent biochemical studies helped define the binding partners of phosphorylated LAT molecules, and showed that in T cells most of the signalling activity of LAT is funnelled through the four C-terminal tyrosine residues found at positions 136, 175, 195 and 235 of the mouse LAT sequence. After TCR-induced phosphorylation, these four tyrosines manifest some specialization in the SH2 domain-containing proteins they recruit. For instance, mutation of tyrosine (Y) 136 primarily eliminates binding of phospholipase Cγ1 (PLCγ1), whereas the simultaneous mutation of Y175 and Y195, or of Y175, Y195 and Y235 results in loss of binding of the Gads and Grb2/Grap adaptors, respectively (Lin & Weiss 2001, Paz et al 2001, Zhang et al 2000, Zhu et al 2003). Gads interacts constitutively with the adaptor SLP76, thereby recruiting it to LAT, together with its constellation of associated molecules (Vav, Nck, Itk, ADAP). SLP-76 contributes to PLCγ1 activation by stabilizing the LAT–PLCγ1 association and by bringing the Tec family PTK Itk in the vicinity of its PLCγ1 substrate (Yablonski et al 2001). In addition to PLCγ1, another major effector molecule functioning downstream of LAT is the Ras GTPase, whose activation is defective in both *Lat*- and *Slp76*-deficient T cells. In T cells, the functional coupling between LAT and Ras occurs mainly through a SLP76–PLCγ1–RasGRP1 pathway, and secondarily via a Grb2–Sos axis.

Mice with a LAT Y136F mutation

To address the importance of LAT Y136 *in vivo*, and the consequence of selectively eliminating binding of PLCγ1, knock-in mice with a mutation that replaced Y136 with phenylalanine (Y136F) were independently derived by two groups (Aguado et al 2002, Sommers et al 2002). Thymuses from mice homozygous for this mutation, Lat^{Y136F}, contained approximately 10-fold fewer cells than wild-type thymuses, and showed reduced numbers of DP and SP thymocytes. Analysis of the DN compartment found in Lat^{Y136F} thymuses further demonstrated that the Lat^{Y136F} mutation constitutes an hypomorphic (partial loss of function) mutation of the pre-TCR checkpoint. Lat^{Y136F} thymi contain very small numbers of CD4 and CD8 SP thymocytes, suggesting that the Lat^{Y136F} mutation also affects the DP to SP transition (Aguadoet al 2002). Therefore, the Lat^{Y136F} mutation negatively affects the two checkpoints that punctuate intrathymic αβ T cell development and globally results in a severe but partial impairment of αβ T cell development.

Given the scarcity of SP thymocytes found in Lat^{Y136F} newborn mice, one would expect very few SP cells in secondary lymphoid organs. However, T cells are readily found in the spleen and lymph nodes of Lat^{Y136F} mice (Aguado et al 2002, Sommers et al 2002). These are primarily an expanding population of $CD4^+$ T cells. As a result, spleen and lymph nodes enlarge, such that by 7 weeks of age, spleen cellularity is approximately 5 times greater than that of wild-type mice. These

$CD4^+$ T cells have a $CD25^-$, $CD44^{high}$, $CD62L^{low}$, $CD69^+$ phenotype resembling activated memory T cells, and express low levels of TCR on their surface, an attribute that may in part account for their inability to proliferate or increase their level of CD69 upon treatment with anti-TCR or anti-TCR plus anti-CD28 antibodies.

$CD4^+$ T cells freshly isolated from Lat^{Y136F} mice expressed sufficient IL4 and IL10 transcripts to allow their detection even without *ex vivo* restimulation (Aguado et al 2002). Upon activation by phorbol-12-myristate-13-acetate (PMA) and ionomycin, IL5, IL13 and IFNγ transcripts were additionally detected, and close to 80% of the $CD4^+$ T cells expressed high levels of intracytoplasmic IL4. Thus, over the first weeks of life and in the absence of deliberate antigenic stimuli, the $CD4^+$ αβ T cells that expanded in Lat^{Y136F} mice deployed a Th2-like effector program.

Secondary lymphoid organs of 6-week old Lat^{Y136F} mice contained 7 to 10 times more B cells than their wild-type counterparts. Most of the B cells found in those enlarged secondary lymphoid organs were highly activated, and contained antibody-producing cells (Aguado et al 2002, Sommers et al 2002). Serum IgG1 and IgE concentrations were elevated approximately 200 times and up to 10 000 times, respectively, compared to wild-type mice. In contrast, the levels of the other Ig isotypes did not differ significantly from those of wild-type serum. In support of the idea of polyclonal hypergammaglobulinaemia G1 and E, the concentrations of both κ and λ light chains were both augmented in the serum of Lat^{Y136F} mice. Given that isotype switching to IgE and IgG1 depends on IL4 and IL13, the overproduction of IgE and IgG1 noted in Lat^{Y136F} mice is probably secondary to the presence of an abnormally high frequency of Th2 effectors. Analysis of lymph nodes from Lat^{Y136F} mice older than 4 weeks also showed the presence of high levels of eosinophils, probably resulting from the IL5 produced by the expanding $CD4^+$ cells. Important lymphocytic infiltrations were also observed in multiple organs, including lung, kidney and liver.

When purified $CD4^+$ T cells from Lat^{Y136F} mice were adoptively transferred into hosts that are both T cell deficient and B cell proficient (as a result of the $Cd3\text{-}\varepsilon^{\Delta5/\Delta5}$ mutation [Malissen et al 1995]), they expanded over time and converted most host B cells into IgE- and IgG1-producing cells (Y. Wang and M. Malissen, unpublished data). Therefore, the Lat^{Y136F} mutation acts primarily at the level of $CD4^+$ T cells, and the development of the hypergammaglobulinaemia E and G1 does not require the expression of LAT^{Y136F} molecules within B cells.

Positive and negative selection in Lat^{Y136F} mice

To explain the presence in Lat^{Y136F} mutant mice of a lymphoproliferative disorder and of autoantibodies against double-stranded DNA and nucleoproteins, it has been hypothesized that this mutation results in a failure to completely eliminate self-reactive T cells by negative selection (Sommers et al 2002). Autoreactive T cells could then escape to the periphery, where they expand and be causative of the autoimmune syndrome. According to this hypothesis, antibodies against DNA and nuclear antigens result from T–B cooperation events involving T cells that specifically react against self-peptide MHC class II complexes expressed by B cells.

In Lat^{Y136F} mice, the inefficient selective process that allow a few DP T cells to reach the CD4 and CD8 SP stages requires the presence of MHC class II and MHC class I molecules, respectively (Aguado et al 2002). To assess whether this MHC-dependent selection process is associated with impaired negative selection, a TCR αβ transgene specific for MHC class II molecules and originally selected in the context of LAT-sufficient mice, was introduced into mice with a $Lat^{Y136F} \times Rag1^{-/-}$ background. Despite the presence of signs of attempted selection, supporting the view that the TCR–LAT^{Y136F} signalling axis was not completely dead and transmitted some signals upon encounter with intrathymic self peptide–MHC complexes, the Lat^{Y136F} mutation prevented both negative and positive selection, and the TCR^{+} DP cells found in these mice remained essentially in a state of non-selection (Y. Wang and M. Malissen, unpublished data). Therefore an MHC class II-restricted αβ TCR originally calibrated in a LAT-proficient context, triggered neither positive nor negative selection when forced to cooperate with LAT molecules that had a crippled signalling ability.

It remains thus possible that the Lat^{Y136F} mutation commensurably altered the sensitivity of DP thymocytes to both positive and negative selection. Accordingly, we would like to suggest that the SP cells that develop in Lat^{Y136F} mice are appropriately calibrated in the context of the crippled LAT molecules. The low intensity signals expected to emanate from the TCR–LAT^{Y136F} axis supporting the selection of only those DP cells expressing TCR whose affinity for self is shifted toward higher values than in a normal, LAT-proficient background. We recently showed that the Lat^{Y136F} mutation promotes T cell-dependent B cell activation leading to germinal centre, memory and plasma cell formation even in an MHC class II-independent manner (Genton et al 2006). Moreover, B cell activation was polyclonal and not antigen-driven because the increase in serum IgG1 and IgE concentrations involved antibodies and autoantibodies equally (Genton et al 2006). This led us to a hypothesis that does not invoke any intrinsic defect in the process of negative selection in Lat^{Y136F} mice. According to that alternative hypothesis, the presence of autoantibodies in Lat^{Y136F} mice might reflect the fact that the Lat^{Y136F} $CD4^{+}$ effector T cells have acquired the ability to help B cells in a TCR-independent, 'quasi-mitogenic' mode, thereby inducing a massive polyclonal B cell activation that is accompanied by the production of autoantibodies among other antibodies.

Is a LAT signalosome pathology taking shape?

Mice having tyrosine to phenylalanine substitution in the three C-terminal tyrosine residues of Lat (called *Lat*$^{Y7/8/9F}$) showed a strict block in αβ T cell development. However, a few γδ T cells develop in these mice and give rise to a polyclonal lymphoproliferative disorder that resembles the one observed in *Lat*^{Y136F} mice (Nunez-Cruz et al 2003). Therefore, in the absence of any intentional immunization, the *Lat*^{Y136F} and *Lat*$^{Y7/8/9F}$ mutations lead in two distinct T cell lineages to the unfolding of a remarkably similar lymphoproliferative disorders. The recurrent features observed in these two models can be summarized as follow:

(1) Both mutations are recessive: they are only detectable after breeding mutant mice to homozygosity or to mice carrying null allele of the *Lat* gene.
(2) Both mutations result in a severe but incomplete impairment of αβ (*Lat*^{Y136F}) or γδ (*Lat*$^{Y7/8/9F}$) T cell development.
(3) The few T cells that reach the periphery of these mutant mice give rise to polyclonal lymphoproliferative disorders involving either αβ (*Lat*^{Y136F}) or γδ (*Lat*$^{Y7/8/9F}$) T cells.
(4) The αβ and γδ T cells expanding in the periphery of *Lat*^{Y136F} and *Lat*$^{Y7/8/9F}$ mice, respectively, had a $CD25^{-}CD44^{high}CD62L^{low}CD69^{+}$ phenotype closely resembling activated-memory T cells.
(5) Paradoxically, the T cells expanding in the periphery of *Lat*^{Y136F} and *Lat*$^{Y7/8/9F}$ mice are largely refractory to direct TCR stimulation *in vitro.*
(6) The T cells expanding in the periphery of *Lat*^{Y136F} and *Lat*$^{Y7/8/9F}$ mice express low levels of TCR on their surface, an attribute that only partially accounts for their inability to proliferate in response to TCR stimulation *in vitro.*
(7) In the absence of deliberate antigenic stimuli, the populations of CD4 αβ and γδ T cells that expand in *Lat*^{Y136F} and *Lat*$^{Y7/8/9F}$ mice, respectively, deploy a Th2-like effector program and trigger Th2-type disorder characterized by hypergammaglobulinaemia E and G1 (*Lat*^{Y136F} and *Lat*$^{Y7/8/9F}$ mice), and tissue eosinophilia (*Lat*^{Y136F} mice).
(8) Autoantibodies against DNA and nuclear antigens are present at least in the serum of *Lat*^{Y136F} mice.
(9) Despite prominent lymphocytic infiltrations in the thymus, lung, liver and kidney, homozygous *Lat*^{Y136F} and *Lat*$^{Y7/8/9F}$ mice showed no chronic intestinal inflammation or tumour formation.

Mice mutated for genes encoding molecules belonging to the LAT signalosome or lying proximal to it (e.g. c-Cbl and RasGRP1 [Chiang et al 2004, Priatel et al 2002]) showed a phenotype that largely recapitulates that of *Lat*^{Y136F} and *Lat*$^{Y7/8/9F}$ mutant mice. Although a careful side-by-side comparison of these various mutant mice remains to be done in the same laboratory, these data strongly suggest the existence of an immunopathology proper to the LAT signalosome. We propose to coin this novel pathology as 'LAT signalling pathology' (LSP). LSP differs from other T cell lymphoproliferative disorders due to defects in CD152 (CTLA4)-function or in CD95–CD95L (Fas–FasL) interactions. Interestingly, LSP is reminiscent of the conditions manifested by some patients suffering from idiopathic hypereosinophilic syndrome (Roufosse et al 1999), raising the possibility that some of the afflicted patients may harbour similar mutations in their LAT signalosome.

Perspectives

Although the effects of the Lat^{Y136F} and $Lat^{Y7/8/9F}$ mutations on the development of αβ and γδ T cells convincingly demonstrate the positive regulatory role played by LAT, the lymphoproliferative disorder observed in the periphery of Lat^{Y136F} and $Lat^{Y7/8/9F}$ mice showed that LAT is more than just an activator and likely plays a negative regulatory role. Consistent with that view, phosphorylated LAT has been shown to recruit inhibitory effectors. For instance, the docking of Gab2 to phosphorylated LAT, occurs through Gads/Grb2, and results in the recruitment of inhibitory molecules such as the SHP2 protein tyrosine phosphatase (Yamasaki et al 2003). Therefore, LAT functions appear to be dual in that it directs the recruitment and activation of positive intracytoplasmic effectors (e.g. PLCg1, Vav, RasGRP1), and also contributes to establish a negative feedback loop that controls the homeostasis of peripheral T cells and their differentiation into effector cells. Based on available biochemical data, it is however difficult to understand how the signals originating from distinctly mutated LAT signalosomes lead, in two distinct T cell lineages and in the absence of any intended immunization, to the unfolding of a conspicuously similar developmental and terminal differentiation program.

It has been recently suggested that the presence of abnormal $CD4^+CD25^+$ regulatory T cells, contribute to the uncontrolled expansion of $CD4^+$ T cells that develops in Lat^{Y136F} mutant mice (Koonpaew et al 2006). However, the pathology that unfolds in Lat^{Y136F} and $Lat^{Y7/8/9F}$ mice is distinct from the pathology that occurs in mice deprived of $CD4^+CD25^+$ regulatory T cells, suggesting that features intrinsic to the Lat^{Y136F} $CD4^+$ or $Lat^{Y7/8/9F}$ γδ T cells likely contribute to the Th2 lymphoproliferative disorder. For instance, the Lat^{Y136F} and $Lat^{Y7/8/9F}$ signalosomes may fail to bind PLCγ1 and to induce calcium flux, leading to a lack of FasL expression and in turn to a failure of apoptosis of the Lat^{Y136F} $CD4^+$ or $Lat^{Y7/8/9F}$ γδ T cells. It will thus be more the absence of signals needed to trigger FasL rather than the lack of recruitment of *bona fide* negative regulators that will account for the LAT signalling pathology. According to that scenario, it has to be inferred that some signals, falling below the threshold needed to trigger FaL, are still generated through the crippled TCR–LAT axis and suffice to trigger the expansion of $CD4^+$

and γδ T cells and their differentiation into Th2 effector cells. It remains also possible that the Th2-type effector T cells expanding in the periphery of Lat^{Y136F} and $Lat^{Y7/8/9F}$ mice are no longer subjected to the control of the TCR–LAT axis. For instance, the absence of signalling competent LAT molecules may relieve the control the TCR normally exerts on the CD28 co-stimulator pathway, and this last pathway may then control the physiology of the expending $CD4^{+}$ T cells.

Acknowledgements

Supported by CNRS, INSERM, ARC, FRM (Défis de la Recherche en Allergie), the European Communities (MUGEN Network of Excellence), and the RIO/MNG Platform.

References

Aguado E, Richelme S, Nuñez-Cruz S et al 2002 Induction of T helper type 2 immunity by a point mutation in the LAT adaptor. Science 296:2036–2040

Chiang YJ, Sommers CL, Jordan MS et al 2004 Inactivation of c-Cbl reverses neonatal lethality and T cell developmental arrest of SLP-76-deficient mice. J Exp Med 200:25–34

Finco TS, Kadlecek T, Zhang W, Samelson LE, Weiss A 1998 LAT is required for TCR-mediated activation of PLCgamma1 and the Ras pathway. Immunity 9:617–627

Genton C, Wang Y, Izui S et al 2006 The Th2 lymphoproliferation developing in LatY136F mutant mice triggers polyclonal B cell activation and systemic autoimmunity. J Immunol 177:2285–2293

Koonpaew S, Shen S, Flowers L, Zhang W 2006 LAT-mediated signaling in CD4+CD25+ regulatory T cell development. J Exp Med 203:119–129

Lin J, Weiss A 2001 Identification of the minimal tyrosine residues required for linker for activation of T cell function. J Biol Chem 276:29588–29595

Malissen M, Gillet A, Ardouin L et al 1995 Altered T cell development in mice with a targeted mutation of the CD3-epsilon gene. EMBO J 14:4641–4653

Nunez-Cruz S, Aguado E, Richelme S et al 2003 LAT regulates gammadelta T cell homeostasis and differentiation. Nat Immunol 4:999–1008

Paz PE, Wang S, Clarke H, Lu X, Stokoe D, Abo A 2001 Mapping the Zap-70 phosphorylation sites on LAT (linker for activation of T cells) required for recruitment and activation of signalling proteins in T cells. Biochem J 356:461–471

Priatel JJ, Teh SJ, Dower NA, Stone JC, Teh HS 2002 RasGRP1 transduces low-grade TCR signals which are critical for T cell development, homeostasis, and differentiation. Immunity 17:617–627

Roufosse F, Schandene L, Sibille C et al 1999 T-cell receptor-independent activation of clonal Th2 cells associated with chronic hypereosinophilia. Blood 94:994–1002

Sommers CL, Park CS, Lee J et al 2002 A LAT mutation that inhibits T cell development yet induces lymphoproliferation. Science 296:2040–2043

Yablonski D, Kadlecek T, Weiss A 2001 Identification of a phospholipase C-gamma1 (PLC-gamma1) SH3 domain-binding site in SLP-76 required for T-cell receptor-mediated activation of PLC-gamma1 and NFAT. Mol Cell Biol 21:4208–4218

Yamasaki S, Nishida K, Sakuma M et al 2003 Gads/Grb2-mediated association with LAT is critical for the inhibitory function of Gab2 in T cells. Mol Cell Biol 23:2515–2529

Zhang W, Sloan-Lancaster J, Kitchen J, Trible RP, Samelson LE 1998 LAT: the ZAP-70 tyrosine kinase substrate that links T cell receptor to cellular activation. Cell 92:83–92

Zhang W, Irvin BJ, Trible RP, Abraham RT, Samelson LE 1999 Functional analysis of LAT in TCR-mediated signaling pathways using a LAT-deficient Jurkat cell line. Int Immunol 11:943–950

Zhang W, Trible RP, Zhu M, Liu SK, McGlade CJ, Samelson LE 2000 Association of Grb2, Gads, and phospholipase C-gamma 1 with phosphorylated LAT tyrosine residues. Effect of LAT tyrosine mutations on T cell antigen receptor-mediated signaling. J Biol Chem 275:23355–23361

Zhu M, Janssen E, Zhang W 2003 Minimal requirement of tyrosine residues of linker for activation of T cells in TCR signaling and thymocyte development. J Immunol 170:325–333

DISCUSSION

Cook: Have you enumerated the regulatory T cells (T_{reg}) in both populations and seen whether they work?

Malissen: Yes. Some colleagues working at NIH have been working on mice similar to us and analysed this issue. They published a recent paper about T_{regs}, and we have divergent results. They enumerated T_{regs} using an anti-Foxp3 antibody and focused on the $CD25^+$ T_{reg} cells. We developed some Foxp3-IRESGFP reporter mice and crossed these mice onto our LAT^{Y136F} mutant mice. In these mutant mice, T_{regs} are present and most of them are $CD25^-$. Therefore there are T_{regs} present, but we still don't know whether they function as well as wild-type T_{regs}. It is probably similar to the CTLA4 knockout where the T_{regs} don't function as well.

Goodnow: Can you cure the mice by giving them sorted wild-type T_{regs} in small numbers?

Malissen: When we were doing the adoptive transfer experiment, initially our intention was to answer that question. Using our Foxp3IRES GFP mouse we sorted out the GFP^+ and $^-$ $CD4^+$ cells. When we gave the GFP^- cells as a control they were also capable of abolition of lymphoproliferative disease. We believe that they are competing for IL7. These cells still respond to environmental cues; they are not transformed cells. If you do the adoptive transfer into a recipient that already has T cells, they don't proliferate.

Lam: In these various LAT transgenic, knock-in and knockout mice, do you see incidence of tumour formation in the T cell compartment?

Malissen: Yes. In some other experiments where we are trying to do some epistatic mapping in the pre-TCR cassette, we have been using the LCK^{Y505F} mutant. When it is put into a wild-type background there is no oncogenesis. In contrast when LCK^{Y505F} is put on a CD45 knockout, a negative loop is removed from the LCK^{Y505F} mutant kinase and the mice die of T lymphoma between 7 and 10 weeks of age. Interestingly, when LCK^{Y505F} is introduced into a LAT-deficient background all the resulting mice die of T lymphoma. This is evidence for negative feedback operated by LAT on LCK. It is amazingly similar to the CD45 situation.

Lam: Are these pre-T cell lymphomas or T cell lymphomas?

Malissen: Essentially, they are both. This shows that LAT is not only a positive regulator, but also a strong negative regulator.

Casanova: You say they do not respond well to antigens, but have you challenged the mice with 'Th1' or 'Th2' pathogens?

Malissen: Yes, the mice were challenged with *Listeria*. They continue to go into a Th2 pathway.

Casanova: Do they control the infection well?

Malissen: No they don't. They just continue to proliferate. In a way, they are behaving in an autistic manner.

Maizels: There are two explanations for the exclusive Th2 polarization through LAT. One would be the strength of signal purely through TCR. Or perhaps there is a separate receptor pathway feeding through LAT. Do you think this might be the case and if so what might the receptors be?

Malissen: It is possible that the TCR–LAT axis does not operate in these cells, and that most of the signals they receive are conveyed through CD28, leading to a Th2 polarization.

Cyster: If an individual had a low expression of LAT would this predispose to Th2 cell development?

Malissen: I believe so.

Hume: LAT is also expressed by NK cells. Is there an NK cell phenotype?

Malissen: Unfortunately not.

Hume: Even in a GVH situation?

Malissen: No.

Hume: In your gene list you emphasized number one, but number two is my favourite growth factor. Do the T cells themselves make macrophage CSF in the LAT mutation?

Malissen: Yes, based on the transcriptome.

Hume: CSF-1 itself causes total immune suppression if you administer it to a mouse, but it is also made as an integral membrane protein. It would actually lead to macrophage T cell interactions that would not normally happen. This could be a signalling pathway to polarization.

Malissen: That is a good point. We are really trying to identify the accessory cells that are talking to the $CD4^+$ T cells

Hume: If you treat a mouse with CSF-1 and take the spleen, it won't stimulate or respond in an MLR. Even if you separate the T cells they won't respond in an MLR. So they receive a signal generated from CSF-1 that polarizes them.

Mackay: Some of the effects on the B cell system are not only through Th2 cells, but also through follicular homing cells and other cytokines such as IL21. A good experiment might be to cross your mice onto a Th2-deficient mouse to see whether there are other effects besides pure IL4. IL21 is like a dynamite stimulator of IgE, for instance.

Malissen: I would like to do that.

Mackay: Carola Vinuesa and I wrote a review on follicular helper cells. The relationship of them to Th2 cells is still uncertain.

Goodnow: Arthur Weiss has shown nicely that LCK activates the SLAP/Cbl pathway to promote endocytosis and keep the TCR from recycling. What do you think might be the relationship to human hyper-IgE syndromes?

Malissen: We have been checking a few patients from Belgium suffering from idiopathic hypereosinophilia but have not found any mutation in LAT.

Lam: Have you looked at the expression of activation markers on T cells, such as ICOS?

Malissen: Yes, they are ICOS positive.

Genetic analysis of systemic autoimmunity

Carola G. Vinuesa and Matthew C. Cook*

*John Curtin School of Medical Research and *Medical School, Australian National University, Canberra, Australia*

Abstract. Even for complex diseases with high rates of monozygotic twin concordance, disease-associated alleles remain elusive. One explanation is that multiple common genetic variants with weak effects cause these diseases and identification of any single allele requires large cohorts. Conversely, if the allelic spectrum of complex disease is heterogeneous, strong effects of rare variants might be offset by their presence in only a small proportion of the patient population. Lupus (SLE) is a systemic autoimmune disease, with significant monozygotic twin concordance, protean clinical manifestations, and production of high-affinity pathogenic autoantibodies. This complex phenotype and results from genome scans point to multiple molecular defects. Contrary to this expectation, our analysis of ENU-mutagenized mice indicates that homozygous mutations frequently cause anti-nuclear antibodies (ANAs), and can account for a full blown lupus phenotype. The best characterized example is the *sanroque* strain, which develops high-affinity dsDNA autoantibodies and fails to censor self-reactive germinal centre T cells. Mapping the underlying mutation identified not only a novel gene, *Roquin*, but also a novel pathogenic pathway for SLE. Identification of such rare variants with strong effects is likely to identify pathogenic pathways that underlie pathology in many patients, lead to interacting molecular partners that also cause pathology, and identify the most effective therapeutic targets.

2007 Decoding the genomic control of immune reactions. Wiley, Chichester (Novartis Foundation Symposium 281) p 103–128

Systemic lupus erythematosus (SLE) is an uncommon but prototypical systemic autoimmune disease (prevalence approximately 1/2000) (Rus & Hochberg 2001). In the childbearing years, females are over-represented by about 9 : 1. The constellation of clinical manifestations present in any individual varies considerably. The more common presentations involve skin, joints, kidney and inflammation at serosal surfaces. The pathophysiological hallmark is production of high-affinity IgG autoantibodies, indicative of peripheral B cell selection in antigen-driven responses within germinal centres, which appear to be produced sometime long before clinical manifestations become apparent (Arbuckle et al 2003). High affinity autoantibodies are not unique to lupus. However, the overall diversity of autoantibodies, together with several key specificities (chromatin, ribonucleoproteins and phospholipids) distinguish lupus from other autoimmune diseases. These key lupus autoantigens are normally confined to the intracellular compartment but are selectively extruded on surface cellular blebs during apoptosis (Casciola-Rosen et al 1996), suggesting that aberrant disposal of apoptotic cells also contributes to lupus pathogenesis. While some end-organ pathology such as glomerulonephritis appears to be a direct consequence of autoantibodies, for other manifestations the mechanism of organ pathology remains obscure.

There is a significant genetic contribution to lupus. The rate of concordance between monozygotic twins has been estimated at 30–70%, approximately 10 times the rate in dizogotic twins (Deapen et al 1992). The risk of lupus in individuals with an affected sibling relative to those without is 20–40 (sibling recurrence risk, λs) (Wandstrat & Wakeland 2001). This indicates a significant genetic component even in comparison with other autoimmune diseases (Wordsworth 1995, Risch 1987).

Identification of lupus loci in humans and mice

Linkage analysis of human multiplex kindreds carried out over the last 20 years has identified numerous lupus-susceptibility loci (Moser et al 1998, Gaffney et al 1998, 2000, Shai et al 1999, Johanneson et al 1999, Cantor et al 2004). Eight cytogenetic locations have been confirmed independently to be in linkage with lupus: 1q23, 1q25-31, 1q41-42, 2q35-37, 4p15.2-16, 6p11-21, 12p24 and 16q12 (Tsao 2004).

As with other complex diseases, cloning disease-associated alleles once linkage regions have been defined has been relatively unsuccessful. However, there is convincing evidence for lupus-associated alleles in three of the linkage regions: *CRP*, *FCGR2A*, *FCGR2B* and *FCGR3A* at 1q23 (Zuniga et al 2001, Kyogoku et al 2002, Manger et al 2002, Lee et al 2002, Magnusson et al 2004), *PARP1* at 1q41-42 (although studies of different *PARP1* alleles have yielded contradictory results) (Tsao et al 1999, Criswell et al 2000), *PDCD1* at 2q37 (Prokunina et al 2002) and the HLA-DR2 C4Q0 haplotype and *TNF* at 6p21 (Graham et al 2002). Also a preliminary report suggests *ZNF423* (OAZ, OLF1/EBF-associated zinc finger protein) could explain the linkage to 16q12 but this needs confirmation. So far, no candidates have been found to explain the linkages observed at 1q25-31, 4p15.2-16 and 12q24.

Several inbred strains of mice exhibit features of lupus, in which more than 45 loci on 17 autosomes cosegregate with the disease phenotype (Jorgensen et al 2004) (33 of them in New Zealand mice). Several human loci are syntenic to mouse lupus loci, including the three loci at chromosome 1. So far there are only three candidate genes, *Cr2*, *Ifi202b* and *C1qa* that may explain the linkages observed at the Sle1c, Nba2 and Nba1 loci, respectively.

Genes that affect B cell activation threshold, disposal of apoptotic cells, and mechanisms that maintain immunological tolerance are plausible autoimmunity candidates. These pathways have already been heavily mined, so numerous candidates are available, and many have been genetically engineered to produce lupus-prone strains. Allelic associations between candidate genes and lupus, which map to regions not previously identified in genome-wide linkage studies, include *PTPN22* (Wu et al 2005, Reddy et al 2005, Orozco et al 2005), *FCGR2A*, *FCGR2B*, *PBX1*, *SPP1* (osteopontin), *IFR3*, *MBL2*, *TYK2*, *IFR5*, *CR1*, *IL10* and *FCRL3*. Most of these genes confer only very modest effects with odds ratios (OR) <2.

Why are lupus susceptibility genes so hard to find?

Phenotypic heterogeneity

One possible explanation for the failure to identify susceptibility genes in complex disease is phenotypic heterogeneity. This is of particular concern in SLE, where the clinical manifestations are remarkably protean. Indeed, for the purposes of clinical and genetic studies, the lupus definition depends on fulfilling any 4 of 11 diagnostic criteria (Tan et al 1982, Hochberg 1997). Clearly, two individuals may meet the criteria for diagnosis while exhibiting completely independent phenotypes.

Genotypic explanations

The genetic architecture of lupus, like all complex diseases, remains unknown. Potential non-mutually exclusive scenarios are first, multiple weak contributions from numerous common alleles, second, multiple contributions from rare alleles, and third, strong contributions from rare alleles (Smith & Lusis 2002, Pritchard & Cox 2002, Reich & Lander 2001). Under the polygenic model, identification of causal alleles is potentially thwarted by weak effects, epistatic interactions, and non-coding mutations. Furthermore, linkage regions often contain more than one candidate. For example, the 1q23 locus contains *FCGR2A*, *FCGR2B*, *FCGR3A* and *PBX1*, as well as the SLAM/CD2 gene cluster of lymphocyte receptors (including *SLAMF1, CD84, CD2, CD244, CD48, LY9, SLAMF6 [LY108]*) and 6p21 contains HLA-DR2 and DR3, plus *C2*, *C4*, *TNF* and *LTA*. If common polymorphisms are in linkage disequilibrium within these loci, determining which of these alleles (one or more) is causal is problematic. This appears to be the case for the *FCGR2A* -R131 and *FCGR3A* F-158 (F176 counting the leader sequence) polymorphisms.

The importance of epistasis in the development of the lupus phenotype has been borne out elegantly with the dissection of murine loci by construction of congenic mice (Haywood et al 2004, Croker et al 2003, Morel et al 2000). For example, on a BL/6 background, neither *Sle1* nor *Yaa* alone are sufficient to cause lupus, whereas *Sle1*/*Yaa* double congenics develop ANAs and fatal glomerulonephritis. Epistatic interactions have not been explored in the genome-wide linkage analyses performed to date. Hap-Map data and the use of dense single nucleotide polymorphism (SNP) arrays might identify multiple gene interactions.

If multiple genes with weak effects mediate complex diseases like lupus, then conservative and non-coding mutations are likely to dominate, since each individual allele will be close to neutral. Two meta-analyses of common variants in common complex diseases conclude that about 50% of candidate causal variants are not rare (minor allele frequency of >0.05) (Ioannidis et al 2001, Lohmueller et al 2003). Analysis of these proven associations reveals that the nature of mutations in complex disease are qualitatively different from those observed in Mendelian disease (Thomas & Kejariwal 2004). Substitutions associated with complex diseases exhibit characteristics similar to those of normal human variation, whereas those in Mendelian diseases are biased significantly towards those that cause non-conservative amino acid changes at highly conserved locations, resulting in a high impact on protein function. This appears to be consistent with the common variant/common disease hypothesis (Reich & Lander 2001), which considers the impact of the human population expansion (approximately 18 000–150 000 years ago) on the allelic spectrum of rare and common disease-related alleles. Assuming a similar mutation rate for all genes, after the population expansion, both rare and common alleles will be replaced at a similar rate by new mutations, but this effect will be proportionately greater on the rare alleles, since they will be present in a smaller population reservoir, and will be diluted more rapidly by a constant rate of new mutations. Since purifying selection is likely to explain why rare disease-associated alleles were rare to begin with, on-going selection is also likely to contribute to further diversification of rare disease-associated alleles.

Alternatively, it has been suggested that the allelic spectrum of complex disease will be similar to the allelic spectrum across the genome, following the predicted distribution of neutral SNPs (Wang & Pike 2004, Ewens 1972). However, if weak purifying selection operates on disease-associated alleles, modelling suggests much greater polymorphism at any particular susceptibility locus (Pritchard 2001). A related question concerns the distribution of phenotypic effects of variant alleles. Based on mutagenesis studies in *Drosophila* and evolutionary theory, it seems likely that an exponential function will relate allele frequency and effect size, with a small number of rare variants exhibiting a strong effect, and a large number of common variants exhibiting a weak effect (Lai et al 1994, Orr 2002, Wang et al 2005).

Allelic spectrum in lupus

Overall, ANAs are much more common than lupus. ANAs may represent a general tendency towards autoimmunity. Almost all lupus patients have ANAs, but ANA alone has a low positive predictive value for the diagnosis of lupus. On the basis of evidence from mouse and human studies, the ANA phenotype appears to have a large genetic footprint. Furthermore, many autoimmunity alleles are associated with more than one autoimmune disease (e.g. *PTPN22*: RA, SLE, T1D autoimmune thyroid disease [Bottini et al 2004]; *FCRL3*: RA, SLE, thyroiditis [Kochi et al 2005], Table 1), and therefore cannot easily account for either the relative specificity of the autoantibody spectrum, or the end-organ pathology that distinguishes lupus from other autoantibody-mediated diseases (e.g. myasthenia gravis, Goodpasture's syndrome).

Wakeland's threshold liability model suggests that the SLE phenotype results from crossing a critical threshold number of active SLE-associated genes (Subramanian & Wakeland 2005). This model is based on observations of the effects of individual murine loci on mouse lupus development. It predicts that 4% of all individuals that simply present with ANAs but no other lupus manifestations have relatively common alleles that promote loss of tolerance to nuclear antigens. In NZ mice these would be represented by *Sle1a*, *Sle1b*, *Sle1c*, etc. A proportion of these individuals may also have alleles that result in dysregulation of the immune system (i.e. *Sle2*, *Sle3*, *Sle5*, *Fas*), and may have some manifestations of lymphoid hyperactivity but still do not present with full blown disease. Alleles that result in end-organ targeting (i.e. *FCGR3A*, *Sle1d*) are necessary for pathogenic autoimmunity, typically with familial aggregation and characteristic end-organ manifestations (nephritis, neurological disorders, arthritis, vasculitis, etc.).

What remains to be determined is whether overall, the allelic spectrum of the general tendency towards autoimmunity, including the production of ANAs, is similar to that of specific end-organ pathology phenotypes that are identified as discrete autoimmune diseases (Fig. 1). In other words, is lupus the result of the cumulative effect of multiple qualitatively similar alleles, or does the added presence of rare alleles explain the pathophysiological pathway that distinguishes lupus from other autoimmune disease? Since weak autoimmunity alleles, which affect thresholds for lymphocyte activation, may confer some selective advantage in the context of pathogenic threat, they might obtain higher population frequencies (Ueda et al 2003). By contrast the complete lupus phenotype is rare and is likely to be subject to purifying selection by virtue of its presentation in women in childbearing years and effects on fecundity. It is then plausible that general autoimmunity alleles and autoimmune disease-specific alleles will follow different distributions.

There are arguments against alleles with strong phenotypic effects in lupus. First, very few single genes in humans have so far been found to confer strong

TABLE 1 Lupus-associated alleles

Gene	*Human*	*Replicated*	*OR*	*Allele*	*Mechanism*	*Other diseases*	*References*
PTPN22	1p13	No	1.37	W620; C1858T	T cell activation	RA, T1D, Graves' disease	Bottini et al 2004, Orozco et al 2005, Reddy et al 2005
C1Q	1p36.3	No		Stop codon	Handling of apoptotic debris	Infection	Manderson et al 2004
FCRL3	1q21	Yes	1.49	−169C	Altered expression and NF-κB binding	RA, AITD	Kochi et al 2005
FCGR2A	1q22–23	Yes	1.3	R131	IC handling; reduced IgG2 binding		Kyogoku et al 2002, Jonsen et al 2004, Magnusson et al 2004
FCGR2B	1q23	Yes		232T	IC handling; B cell signalling theshold; IgG2 binding; plasma cell differentiation		Kono et al 2005
FCGR3A	1q23	Yes	1.6	F158	IC handling; reduced IgG1/3 binding		Karassa et al 2003, Magnusson et al 2004
FASLG	1q23	Yes	$P = 0.024$	−844C	Tolerance	ALPS	Wu et al 1996
CRP	1q23.2	Yes		CRP4 (GT intronic repeat)	Handling of apoptotic debris	Vascular disease	Russell et al 2004
CTLA4	2q33	No	1.3	C-318T; A49G; 3′(AT)n	T cell activation threshold; tolerance	T1D; RA; MG; CD	Ueda et al 2003
PDCD1	2q35–37	Yes	1.3–2.0	PD-1.3A	T cell signalling theshold tolerance	T1D; RA; Allergy	Prokunina et al 2002

SPP1	4q22	No	2.35	−156; 1.57 +1239	Macrophage and B cell activation threshold	RA	Forton et al 2002
HLA DR3, DR2	6p11–22	Yes	2.8	HLA DR3-DQ2-C4AQ0	Antigen presentation		Graham et al 2002, Schur et al 1982
C2	6p21	Yes		A25, B18, Drw2, BFS, C2Q0, C4A4B2 (type I)	IC handling; B cell activation threshold	HSP; polymyositis	Manderson et al 2004, Meyer et al 1985
C4A	6p21	Yes		AQ0	IC handling	T1D; bacterial meningitis	Meyer et al 1985
TNF	6p21.3	Yes	1.6	−308A	End-organ damage	Cerebral malaria susceptibility; mucocutaneous leishmaniasis; scarring trachomata; lepromatous leprosy; possible RA, asthma, AS	Sigurdsson et al 2005
IRF5	7q31	No	1.6	rs2004640 intronic SNP	End-organ damage: IFNα response		Sigurdsson et al 2005
MBL2	10q21	Yes	1.406	54A 57A LX	IC handling	Susceptibility to infection	Lee et al 2005, Sullivan et al 1996
FAS	10q24	No	RR 5.0	297C; 416G	Tolerance	ALPS	Horiuchi et al 1999
TYK2	19p13	No	1.6	S684	End-organ damage: IFNα response	RA	Sigurdsson et al 2005

AITD, autoimmune thyroid disease; ALPS, autoimmune lymphoproliferative syndrome; AS, ankylosing spondylitis; CD, celiac disease; HSP, Henoch-Schonlein purpura; MG, myasthenia gravis; RA, rheumatoid arthritis; T1D, type 1 diabetes.

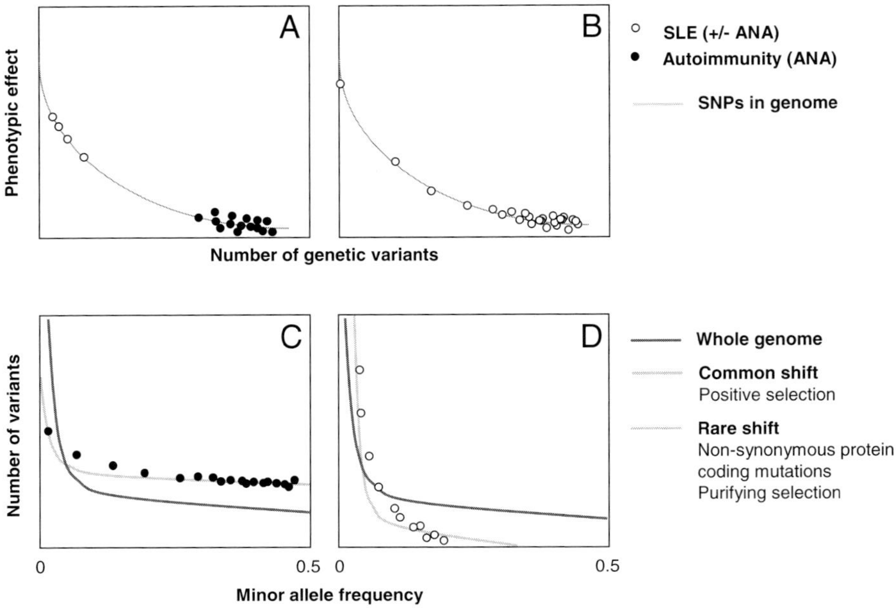

FIG. 1. Scenarios to describe the allelic architecture of lupus. (A,B) Distribution of phenotypic effects of lupus and ANA associated alleles relative to the expected distribution of all SNPs (grey line). Lupus may be the outcome of the composite effect of numerous alleles that each cause a weak tendency towards autoimmunity (ANA production) (filled circles). Alternatively, an additional distinct set of alleles may account for the discrete lupus phenotype (open circles). A conservative estimate is that lupus-associated alleles will follow the overall QTL distribution (A and B), with the majority associated with weak phenotypic effects. (C,D) The allelic spectrum of ANA and lupus genes relative to variation across the genome (dark grey line). ANA is a common phenotype, and associated alleles may offer a weak selective advantage, resulting in a common shift (C) whereas lupus-associated alleles are likely to be rare and subject to purifying selection (D).

risk for SLE development, although this may be a consequence of the methods used to identify these genes. Second, many genetically-manipulated mouse models that identified single genes leading to murine SLE have now been questioned due to confounding genetic background or strain-dependent effects. For example, the effects of *Fas*, *Fcgr2b*, *C1qa*, *Apcs* (SAP) and *Pdcd1* turn out to reflect, at least in part, passively transferred lupus susceptibility genes from the 129/Sv ES cells used to generate the knockouts (Bygrave et al 2004). In the case of *Apcs* and *Fcgr2b*, the 129 interval surrounding these genes is sufficient to generate ANAs (including dsDNA on a BL/6 background).

On the other hand, evidence for genetic heterogeneity in lupus includes identification of linkage regions after stratification of pedigrees according to specific clinical manifestations, which was not previously evident. Examples include thrombocytopenia (where the LOD score at 1q22-23 increased from 2.75 to 3.71), the presence of a relative with self-reported rheumatoid arthritis (RA) (which revealed a novel linkage to 5p 15.3) (Namjoy et al 2002), neuropsychiatric disease (Nath et al 2002), anti-dsDNA antibodies (Namjou et al 2002), haemolytic anaemia, nucleolar ANA, renal disease (Quintero-Del-Rio et al 2002), vitiligo and discoid lupus erythematosus (Nath et al 2001). Finally, C1q represents an important precedent for a rare allele with strong effect resulting in the complex lupus phenotype. Multiple rare mutations that result in C1q deficiency cause severe, early onset lupus in more than 80% of affected individuals, and the penetrance of lupus characterized by ANA, dsDNA, skin, kidney and CNS disease is >90%, with a sex ratio of close to one (Meyer et al 1985, Manderson et al 2004).

Using ENU to discover lupus genes

By virtue of its high rate of random mutagenesis, and the overwhelming predominance of point mutations (Russell et al 1979, Vinuesa & Goodnow 2004), ENU is an excellent tool for investigating the allelic spectrum of disease. We have screened G3 mice from ENU-treated founders for the presence of ANA and commenced secondary analysis of mice for additional features of the lupus phenotype. So far, we have identified one strain, *sanroque*, in which a single coding SNP in the *Rc3h1 (Roquin)* gene causes an unequivocal lupus phenotype with ANA, dsDNA antibodies, autoimmune cytopenia, lymphadenopathy and glomerulonephritis in 100% of homozygous pure B6 mice (Vinuesa et al 2005). Human *RC3H1* is located in 1q25, within one of the 8 linkage regions that have been confirmed, and no candidate gene has been identified in this region so far. It only lies 10 Mb downstream from another linked region, 1q23, for which only SLE susceptibility genes with weak effects have been identified. Thus, *Rc3h1* emerges as a very attractive gene to explain the linkage identified at either of the 1q23 or 1q25 loci.

ENU colony analysis and lessons learnt

ENU mimics natural variation introducing single nucleotide changes spread throughout the genome, thus the chances of hitting ANA or lupus susceptibility genes are *a priori* similar to those variations occurring through evolution, with the caveat that inbred strains are closely related, and many alleles only exist in forms present in a small group of mice that gave rise to these strains, so some gain of function mutations with a selective advantage will not be surveyed.

We have surveyed ENU-treated mice on three different genetic backgrounds for ANA: C57BL/6, B10.BR backcrossed to B6, and B6 backcrossed to CBA. The background incidence of ANAs in non ENU mutagenized mice is <2% in all strains. Overall, ~20% of 347 ENU pedigrees screened show clustering of ANAs (at least 50% penetrance for autosomal recessive traits). Significantly, this percentage is very similar on the three genetic backgrounds, despite known differences in lupus predisposition of B6 and CBA genomes. Nevertheless, once an autoimmune (generally ANA+) trait has been identified on a particular genetic background, the likelihood of persisting on a different genetic background is small. For example, only 2 of ~20 ANA+ B6 strains generated ANA+ mice when outcrossed to CBA, which has devastating implications at the time of mapping. Of the ANA+ mice that have generated offspring, simple Mendelian inheritance of the ANA phenotype has occurred in >75% (frequently with higher penetrance in females). There are a few important conclusions that can be drawn from these observations:

(1) A large number of genes (~120) in each individual can contribute to systemic autoimmunity. Given that the specific locus mutation frequency of ENU is 1 in 1000, and assuming there are ~30 000 genes in the mouse genome, each individual G1 pedigree founder mouse will carry about 30 mutagenized genes. Provided sufficient G3 mice are screened, there is the potential to identify 30 loss-of-function mutations that give rise to a recessive phenotype. A frequency of 1 : 5 pedigrees producing ANAs, of which 50–75% are heritable means that alteration of 1 in every 225–300 genes may give rise to production of ANAs. This number may need to be adjusted downwards, since some genes will never be detected due to embryonic lethality, and some appear to be hit more frequently than average. Nevertheless, a reasonable adjusted estimate is that up to 100–150 genes could contribute to autoimmunity. This is not unreasonable, since ~125 genes are known to be important for T cell development, and T cell lymphopaenia is one of at least 10 different pathways that may lead to autoantibody production (Plotz 2003).

(2) The genes contributing to ANA development are likely to be conserved genes important for mounting protective immune responses. We have observed a similar prevalence of ANAs in different mouse strains after ENU mutagenesis. However, the ANA phenotype is often lost in the outcross. This suggests that many alterations in immune function can lead to ANA production. In other words, immune responses operate on the verge of autoimmunity, although these are frequently suppressed by epistatic interactions. Epistatic interactions have important implications for mapping ANA-associated genes. So far, <10% of mutations identified in B6 or B6×B10BR background have been penetrant on a CBA background, indicating suppressor CBA alleles. Given the limited number of polymorphic markers available between C57BL/6 and C57BL/10 strains, our preferred strategy now is to generate pedi-

grees such that G3 mice are on a mixed CBA×B6 background (Fig. 2). This strategy has yielded a comparable 20% incidence in ANA producing pedigrees (Fig. 3, left panel).

(3) A small number of single genes with strong effects lead to SLE. Only a small proportion of ANA+ strains studied so far develops overt clinical autoimmunity. This supports the notion that few single genes lead to full blown lupus phenotypes.

(4) Mendelian inheritance of ANAs. Most ANA-inducing ENU mutations show autosomal dominant or recessive patterns of inheritance, with frequent complete penetrance at least in female mice. This suggests that although epistatic gene effects are likely to be important for full SLE development, single gene defects will mostly account for ANA production.

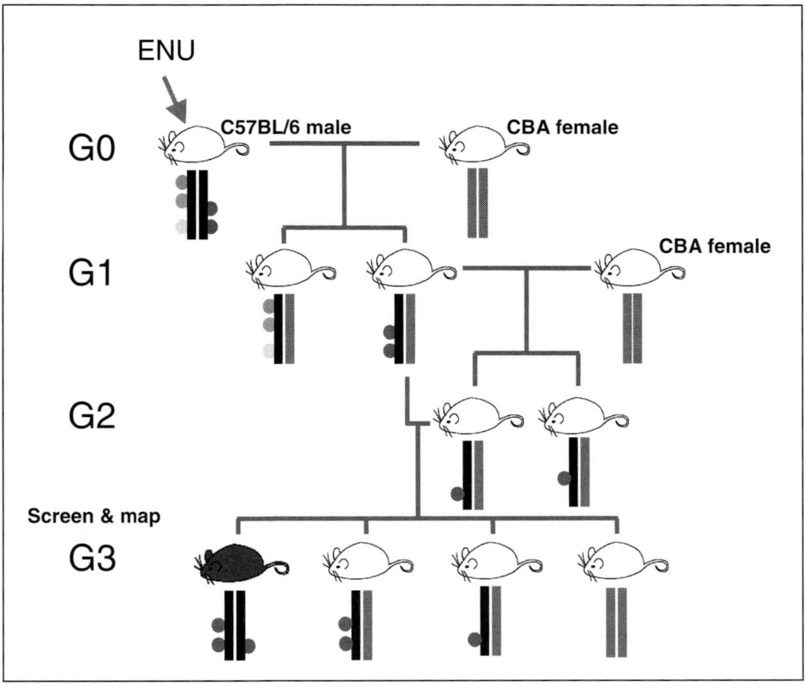

FIG. 2. Strategy for production of G3 mice from ENU-treated founders. ENU-treated B6 mice are outcrossed to CBA to produce G1 mice. G1 (F1) mice are crossed with CBA to produce G2s, which are then backcrossed with G1 parents to produce G3s. G3 are screened for the presence of ANAs. Mice exhibiting this phenotype are scanned for regions of B6 homozygosity to identify a rough map position of the ANA-conferring allele.

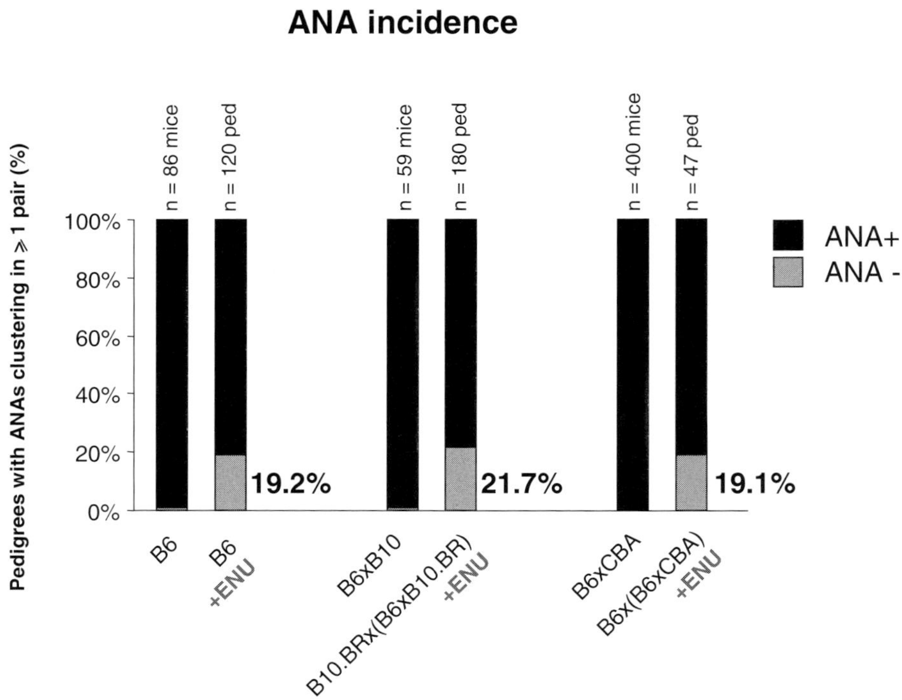

FIG. 3. Incidence of ANAs in ENU strains. C57BL/6 (B6) mice were treated with ENU, from which G3 mice were propagated on three different backgrounds: pure B6 (left columns); B6, B10, B10.BR (middle columns); and B6 CBA (right columns). The incidence of ANA in each of these mutagenized strains is shown along with the background incidence in untreated controls of the same background.

(5) Advantages of carrying out a lupus suppressor screen. Single genes that exert a strong autoimmune effect (e.g. *Roquin*) and are key master regulators of an immune pathway, are likely to play a role also in protective immune responses and thus, not ideal targets for blockade or therapeutic manipulation. Mechanisms controlling end-organ damage may be more amenable to therapeutic intervention, and these pathways can be mined via suppressor screens.

Significance of identifying rare variants

So far, many of the lupus-associated alleles identified exert weak effects (Table 1). Based on knowledge of the biological function of normal alleles (e.g. B cell activation threshold; tendency towards reduced central or peripheral tolerance), disease-associated variants often appear to be consistent with causation but are of limited heuristic value, failing to shed light on the pattern of autoantibody production or end-organ pathology specific for a particular disease. Pathophysiological insights might be obtained from mouse models containing disease-associated alleles, where immune function can be investigated in great detail experimentally. Dissecting epistasis remains a significant obstacle in the dissection of these weak effects (Sullivan et al 2003).

This contrasts dramatically with the outcome from identifying allelic associations of Mendelian diseases, which have frequently revealed a novel pathophysiological pathway. Similarly, important pathophysiological insights have also been obtained from rare Mendelian counterparts of common complex diseases (Bennett et al 2001, Nagamine et al 1997, Yamagata et al 1996, Fajan et al 2001).

Taking into account evidence from mapping studies and our ENU survey, numerous alleles exert weak effects that predispose to autoimmunity/ANA production. Based on the ENU survey, it is likely that many more ANA-associated alleles remain to be identified in humans. Low penetrance after outcrossing mice harbouring these alleles indicates that many depend on epistatic interactions to produce an observable phenotype. This is also consistent with empirical evidence for additive effects of alleles within the same pathway in human lupus (Sullivan et al 2003).

Physiological processes are robust, that is, they are resistant to perturbation, from intrinsic changes (i.e. wild-type phenotype dominance) and environment stimuli (Bhalla & Iyengar 1999). Since the molecular networks underpinning immunity evolved under pressure from pathogens it is likely that pathophysiological pathways of autoimmunity represent perturbations of specific adaptive pathogen-specific responses evoked by environmental stimuli. There is a disproportionate incidence of polymorphisms in genes involved in host defence (Shields et al 1996, Murphy 1993), as well as considerable variation in the prevalence of polymorphisms in autoimmune-associated alleles in different racial groups (Yamada & Ymamoto 2005), consistent with evolution of many solutions to host-defence mechanisms, and consequently, many pathways to autoimmunity. Loss of robustness to the activating stimuli, which modelling suggests could result from multiple variant alleles, or variation in one key molecule, would lead to a maladaptive outcome. In the case of immune signalling networks, this would be predicted to manifest as either aberrant inflammation, susceptibility to infection, or both. The loss of ANA phenotype in ENU-generated mutants after introduction of a different genetic background reflects the robustness of host-defence networks. However, the remarkable similarity of ANA predisposition after ENU-induced variation on several genetic backgrounds suggests that host defence networks operate on the cusp of autoimmunity.

In the short to medium term, therapeutic advances for autoimmune diseases such as lupus will depend on targeting the effector phase of the immune response.

Identification of alleles with strong phenotypic effects should lead to characterization of signature effector responses, since robust pathways of immune responses have been subject to intense selection. The validity of this approach is revealed by the dramatic therapeutic implications of identifying TNF as a key effector in rheumatoid arthritis (Feldmann et al 2004). In the longer term, the availability of more extensive genotyping may enable targeting the minor perturbations that underpin a common clinical phenotype that can be arrived at via many different routes.

References

Arbuckle MR, McClain MT, Rubertone MV et al 2003 Development of autoantibodies before the clinical onset of systemic lupus erythematosus. N Engl J Med 349:1526–1533

Bennett CL, Christie J, Ramsdell F et al 2001 The immune dysregulation, polyendocrinopathy, enteropathy, X-linked syndrome (IPEX) is caused by mutations of FOXP3. Nat Genet 27:20–21

Bhalla US, Iyengar R 1999 Emergent properties of networks of biological signaling pathways. Science 283:381–387

Bottini N, Musumeci L, Alonso A et al 2004 A functional variant of lymphoid tyrosine phosphatase is associated with type I diabetes. Nat Genet 36:337–338

Bygrave AE, Rose KL, Cortes-Hernandez J et al 2004 Spontaneous autoimmunity in 129 and C57BL/6 mice-implications for autoimmunity described in gene-targeted mice. PLoS Biol 2:E243

Cantor RM, Yuan J, Napier S et al 2004 Systemic lupus erythematosus genome scan: support for linkage at 1q23, 2q33, 16q12-13, and 17q21-23 and novel evidence at 3p24, 10q23-24, 13q32, and 18q22-23. Arthritis Rheum 50:3203–3210

Casciola-Rosen L, Rosen A, Petri M, Schlissel M 1996 Surface blebs on apoptotic cells are sites of enhanced procoagulant activity: implications for coagulation events and antigenic spread in systemic lupus erythematosus. Proc Natl Acad Sci USA 93:1624–1629

Criswell LA, Moser KL, Gaffney PM et al 2000 PARP alleles and SLE: failure to confirm association with disease susceptibility. J Clin Invest 105:1501–1502

Croker BP, Gilkeson G, Morel L 2003 Genetic interactions between susceptibility loci reveal epistatic pathogenic networks in murine lupus. Genes Immun 4:575–585

Deapen D, Escalante A, Weinrib L et al 1992 A revised estimate of twin concordance in systemic lupus erythematosus. Arthritis Rheum 35:311–318

Ewens WJ 1972 The sampling theory of selectively neutral alleles. Theor Popul Biol 3:87–112

Fajans SS, Bell GI, Polonsky KS 2001 Molecular mechanisms and clinical pathophysiology of maturity-onset diabetes of the young. N Engl J Med 345:971–980

Feldmann M, Brennan FM, Williams RO, Woody JN, Maini RN 2004 The transfer of a laboratory based hypothesis to a clinically useful therapy: the development of anti-TNF therapy of rheumatoid arthritis. Best Pract Res Clin Rheumatol 18:59–80

Forton AC, Petri MA, Goldman D, Sullivan KE 2002 An osteopontin (SPP1) polymorphism is associated with systemic lupus erythematosus. Hum Mutat 19:459

Gaffney PM, Kearns GM, Shark KB et al 1998 A genome-wide search for susceptibility genes in human systemic lupus erythematosus sib-pair families. Proc Natl Acad Sci USA 95: 14875–14879

Gaffney PM, Ortmann WA, Selby SA et al 2000 Genome screening in human systemic lupus erythematosus: results from a second Minnesota cohort and combined analyses of 187 sib-pair families. Am J Hum Genet 66:547–556

Graham RR, Ortmann WA, Langefeld CD et al 2002 Visualizing human leukocyte antigen class II risk haplotypes in human systemic lupus erythematosus. Am J Hum Genet 71:543–553

Haywood ME, Rogers NJ, Rose SJ et al 2004 Dissection of BXSB lupus phenotype using mice congenic for chromosome 1 demonstrates that separate intervals direct different aspects of disease. J Immunol 173:4277–4285

Hochberg MC 1997 Updating the American College of Rheumatology revised criteria for the classification of systemic lupus erythematosus. Arthritis Rheum 40:1725

Horiuchi T, Nishizaka H, Yasunaga S et al 1999 Association of Fas/APO-1 gene polymorphism with systemic lupus erythematosus in Japanese. Rheumatology (Oxford) 38:516–520

Ioannidis JP, Ntzani EE, Trikalinos TA, Contopoulos-Ioannidis DG 2001 Replication validity of genetic association studies. Nat Genet 29:306–309

Johanneson B, Steinsson K, Lindqvist AK et al 1999 A comparison of genome-scans performed in multicase families with systemic lupus erythematosus from different population groups. J Autoimmun 13:137–141

Jonsen A, Bengtsson AA, Sturfelt G, Truedsson L 2004 Analysis of HLA DR, HLA DQ, C4A, FcgammaRIIa, FcgammaRIIIa, MBL, and IL-1Ra allelic variants in Caucasian systemic lupus erythematosus patients suggests an effect of the combined FcgammaRIIa R/R and IL-1Ra 2/2 genotypes on disease susceptibility. Arthritis Res Ther 6:R557–562

Jorgensen TN, Gubbels MR, Kotzin BL 2004 New insights into disease pathogenesis from mouse lupus genetics. Curr Opin Immunol 16:787–793

Karassa FB, Trikalinos TA, Ioannidis JP 2003 The Fc gamma RIIIA-F158 allele is a risk factor for the development of lupus nephritis: a meta-analysis. Kidney Int 63:1475–1482

Kochi Y, Yamada R, Suzuki A et al 2005 A functional variant in FCRL3, encoding Fc receptor-like 3, is associated with rheumatoid arthritis and several autoimmunities. Nat Genet 37:478–485

Kono H, Kyogoku C, Suzuki T et al 2005 FcgammaRIIB Ile232Thr transmembrane polymorphism associated with human systemic lupus erythematosus decreases affinity to lipid rafts and attenuates inhibitory effects on B cell receptor signaling. Hum Mol Genet 14:2881–2892

Kyogoku C, Dijstelbloem HM, Tsuchiya N et al 2002 Fcgamma receptor gene polymorphisms in Japanese patients with systemic lupus erythematosus: contribution of FCGR2B to genetic susceptibility. Arthritis Rheum 46:1242–1254

Kyogoku C, Tsuchiya N, Matsuta K, Tokunaga K 2002 Studies on the association of Fc gamma receptor IIA, IIB, IIIA and IIIB polymorphisms with rheumatoid arthritis in the Japanese: evidence for a genetic interaction between HLA-DRB1 and FCGR3A. Genes Immun 3: 488–493

Lai C, Lyman RF, Long AD, Langley CH, Mackay TF 1994 Naturally occurring variation in bristle number and DNA polymorphisms at the scabrous locus of Drosophila melanogaster. Science 266:1697–1702

Lee EB, Lee YJ, Baek HJ et al 2002 Fcgamma receptor IIIA polymorphism in Korean patients with systemic lupus erythematosus. Rheumatol Int 21:222–226

Lee YH, Witte T, Momot T et al 2005 The mannose-binding lectin gene polymorphisms and systemic lupus erythematosus: two case-control studies and a meta-analysis. Arthritis Rheum 52:3966–3974

Lohmueller KE, Pearce CL, Pike M, Lander ES, Hirschhorn JN 2003 Meta-analysis of genetic association studies supports a contribution of common variants to susceptibility to common disease. Nat Genet 33:177–182

Magnusson V, Johanneson B, Lima G, Odeberg J, Alarcon-Segovia D, Alarcon-Riquelme ME 2004 Both risk alleles for FcgammaRIIA and FcgammaRIIIA are susceptibility factors for SLE: a unifying hypothesis. Genes Immun 5:130–137

Manderson AP, Botto M, Walport MJ 2004 The role of complement in the development of systemic lupus erythematosus. Annu Rev Immunol 22:431–456

Manger K, Repp R, Jansen M et al 2002 Fcgamma receptor IIa, IIIa, and IIIb polymorphisms in German patients with systemic lupus erythematosus: association with clinical symptoms. Ann Rheum Dis 61:786–792

Meyer O, Hauptmann G, Tappeiner G, Ochs HD, Mascart-Lemone F 1985 Genetic deficiency of C4, C2 or C1q and lupus syndromes. Association with anti-Ro (SS-A) antibodies. Clin Exp Immunol 62:678–684

Morel L, Croker BP, Blenman KR et al 2000 Genetic reconstitution of systemic lupus erythematosus immunopathology with polycongenic murine strains. Proc Natl Acad Sci USA 97:6670–6675

Moser KL, Neas BR, Salmon JE et al 1998 Genome scan of human systemic lupus erythematosus: evidence for linkage on chromosome 1q in African-American pedigrees. Proc Natl Acad Sci USA 95:14869–14874

Murphy PM 1993 Molecular mimicry and the generation of host defense protein diversity. Cell 72:823–826

Nagamine K, Peterson P, Scott HS et al 1997 Positional cloning of the APECED gene. Nat Genet 17:393–398

Namjou B, Nath SK, Kilpatrick J et al 2002 Genome scan stratified by the presence of anti-double-stranded DNA (dsDNA) autoantibody in pedigrees multiplex for systemic lupus erythematosus (SLE) establishes linkages at 19p13.2 (SLED1) and 18q21.1 (SLED2). Genes Immun 3 Suppl 1, S35–41

Namjou B, Nath SK, Kilpatrick J et al 2002 Stratification of pedigrees multiplex for systemic lupus erythematosus and for self-reported rheumatoid arthritis detects a systemic lupus erythematosus susceptibility gene (SLER1) at 5p15.3. Arthritis Rheum 46:2937–2945

Nath SK, Kelly JA, Namjou B et al 2001 Evidence for a susceptibility gene, SLEV1, on chromosome 17p13 in families with vitiligo-related systemic lupus erythematosus. Am J Hum Genet 69:1401–1406

Nath SK, Kelly JA, Reid J et al 2002 SLEB3 in systemic lupus erythematosus (SLE) is strongly related to SLE families ascertained through neuropsychiatric manifestations. Hum Genet 111:54–58

Orozco G, Sanchez E, Gonzalez-Gay MA et al 2005 Association of a functional single-nucleotide polymorphism of PTPN22, encoding lymphoid protein phosphatase, with rheumatoid arthritis and systemic lupus erythematosus. Arthritis Rheum 52:219–224

Orr HA 2002 The population genetics of adaptation: the adaptation of DNA sequences. Evolution Int J Org Evolution 56:1317–1330

Plotz PH 2003 The autoantibody repertoire: searching for order. Nat Rev Immunol 3: 73–78

Pritchard JK 2001 Are rare variants responsible for susceptibility to complex diseases? Am J Hum Genet 69:124–137

Pritchard JK, Cox NJ 2002 The allelic architecture of human disease genes: common disease-common variant . . . or not? Hum Mol Genet 11:2417–2423

Prokunina L, Castillejo-Lopez C, Oberg F et al 2002 A regulatory polymorphism in PDCD1 is associated with susceptibility to systemic lupus erythematosus in humans. Nat Genet 32:666–669

Quintero-Del-Rio AI, Kelly JA, Kilpatrick J, James JA, Harley JB 2002 The genetics of systemic lupus erythematosus stratified by renal disease: linkage at 10q22.3 (SLEN1), 2q34-35 (SLEN2), and 11p15.6 (SLEN3). Genes Immun 3 Suppl 1, S57–62

Reddy MV, Johansson M, Sturfelt G et al 2005 The R620W C/T polymorphism of the gene PTPN22 is associated with SLE independently of the association of PDCD1. Genes Immun 6:658–662

Reich DE, Lander ES 2001 On the allelic spectrum of human disease. Trends Genet 17: 502–510

Risch N 1987 Assessing the role of HLA-linked and unlinked determinants of disease. Am J Hum Genet 40:1–14

Rus VHA, Hochberg MC (eds) 2001 Systemic lupus erythematosus. Oxford University Press, 123–140

Russell AI, Cunninghame Graham DS, Shepherd C et al 2004 Polymorphism at the C-reactive protein locus influences gene expression and predisposes to systemic lupus erythematosus. Hum Mol Genet 13:137–147

Russell WL, Kelly EM, Hunsicker PR, Bangham JW, Maddux SC, Phipps EL 1979 Specific-locus test shows ethylnitrosourea to be the most potent mutagen in the mouse. Proc Natl Acad Sci USA 76:5818–5819

Schur PH, Meyer I, Garovoy M, Carpenter CB 1982 Associations between systemic lupus erythematosus and the major histocompatibility complex: clinical and immunological considerations. Clin Immunol Immunopathol 24:263–275

Shai R, Quismorio FP Jr, Li L et al 1999 Genome-wide screen for systemic lupus erythematosus susceptibility genes in multiplex families. Hum Mol Genet 8:639–644

Shields DC, Harmon DL, Whitehead AS 1996 Evolution of hemopoietic ligands and their receptors. Influence of positive selection on correlated replacements throughout ligand and receptor proteins. J Immunol 156:1062–1070

Sigurdsson S, Nordmark G, Goring HH et al 2005 Polymorphisms in the tyrosine kinase 2 and interferon regulatory factor 5 genes are associated with systemic lupus erythematosus. Am J Hum Genet 76:528–537

Smith DJ, Lusis AJ 2002 The allelic structure of common disease. Hum Mol Genet 11: 2455–2461

Subramanian S, Wakeland EK 2005 The importance of epistatic interactions in the development of autoimmunity. In: Genetics of autoimmunity. Wiley, Chichester (Novartis Found Symp 267) p 76–93

Sullivan KE, Jawad AF, Piliero LM et al 2003 Analysis of polymorphisms affecting immune complex handling in systemic lupus erythematosus. Rheumatology (Oxford) 42:446–452

Sullivan KE, Wooten C, Goldman D, Petri M 1996 Mannose-binding protein genetic polymorphisms in black patients with systemic lupus erythematosus. Arthritis Rheum 39:2046–2051

Tan EM, Cohen AS, Fries JF et al 1982 The 1982 revised criteria for the classification of systemic lupus erythematosus. Arthritis Rheum 25:1271–1277

Thomas PD, Kejariwal A 2004 Coding single-nucleotide polymorphisms associated with complex vs. Mendelian disease: evolutionary evidence for differences in molecular effects. Proc Natl Acad Sci USA 101:15398–15403

Tsao BP 2004 Update on human systemic lupus erythematosus genetics. Curr Opin Rheumatol 16:513–521

Tsao BP, Cantor RM, Grossman JM et al 1999 PARP alleles within the linked chromosomal region are associated with systemic lupus erythematosus. J Clin Invest 103:1135–1140

Ueda H, Howson JM, Esposito L et al 2003 Association of the T-cell regulatory gene CTLA4 with susceptibility to autoimmune disease. Nature 423:506–511

Vinuesa CG, Cook MC, Angelucci C et al 2005 A RING-type ubiquitin ligase family member required to repress follicular helper T cells and autoimmunity. Nature 435:452–458

Vinuesa CG, Goodnow CC 2004 Illuminating autoimmune regulators through controlled variation of the mouse genome sequence. Immunity 20:669–679

Wandstrat A, Wakeland E 2001 The genetics of complex autoimmune diseases: non-MHC susceptibility genes. Nat Immunol 2:802–809

Wang WY, Barratt BJ, Clayton DG, Todd JA 2005 Genome-wide association studies: theoretical and practical concerns. Nat Rev Genet 6 109–118

Wang WY, Pike N 2004 The allelic spectra of common diseases may resemble the allelic spectrum of the full genome. Med Hypotheses 63:748–751
Wordsworth P 1995 Genes and arthritis. Br Med Bull 51:249–266
Wu H, Cantor RM, Graham DS et al 2005 Association analysis of the R620W polymorphism of protein tyrosine phosphatase PTPN22 in systemic lupus erythematosus families: increased T allele frequency in systemic lupus erythematosus patients with autoimmune thyroid disease. Arthritis Rheum 52:2396–2402
Wu J, Wilson J, He J, Xiang L, Schur PH, Mountz JD 1996 Fas ligand mutation in a patient with systemic lupus erythematosus and lymphoproliferative disease. J Clin Invest 98: 1107–1113
Yamada R, Ymamoto K 2005 Recent findings on genes associated with inflammatory disease. Mutat Res 573:136–151
Yamagata K, Oda N, Kaisaki PJ et al 1996 Mutations in the hepatocyte nuclear factor-1alpha gene in maturity-onset diabetes of the young (MODY3). Nature 384:455–458
Zuniga R, Ng S, Peterson MG et al 2001 Low-binding alleles of Fcgamma receptor types IIA and IIIA are inherited independently and are associated with systemic lupus erythematosus in Hispanic patients. Arthritis Rheum 44:361–367

DISCUSSION

Casanova: I think it is great news for the field that you have found Mendelian cases of lupus. There was a case of complement deficiency that you noted, but I think there are a few cases of patients with chronic granulomatous disease and lupus in the literature. Perhaps that is something you could add to your list.

Lam: In your Sanroque mouse B cells do you see any changes in the levels of ICOS ligands?

Vinuesa: We don't, but we do see some changes in other costimulatory molecules.

Lam: What about the levels of Bcl-6?

Vinuesa: We haven't looked at this.

Malissen: Do you know the phenotype of the Roquin knockout?

Vinuesa: We are in the process of making this. We hope to receive the mice soon.

Malissen: Has anyone tried to force the expression of ICOS, to produce a constitutively high level? In the relationship between Roquin and ICOS are there other molecules contributing to the phenotype?

Vinuesa: That is a good question. In the literature there is only an ICOS ligand Fc fusion protein transgenic mouse. These mice have a phenotype of lymphadenopathy, splenomegaly and hypergammaglobulinaemia. We are now trying to do exactly what you are suggesting just with ICOS on the T cells so we can examine the autoimmune phenotype. This is relatively tricky, because we are using retroviral overexpression and we have to activate the T cells before putting them into the mice. This introduces biases.

Malissen: So your hypothesis would be that Roquin has something to do with mRNA turnover?

Vinuesa: Yes. At least one of the things it is doing is to regulate mRNA expression. Whether or not one of the targets for mRNA regulation is ICOS is something that we are looking at at the moment. There is some evidence that ICOS mRNA is increased.

Malissen: You mentioned all these mRNAs that were overexpressed in Sanroque mice. Do they have any kind of sequence identity?

Vinuesa: Good question. We asked an Oxford collaborator to look at this for us. He couldn't see any obvious patterns. Some of these mRNAs do have increased AU-rich elements in their 3' UTRs, and ICOS is one of them, along with IL21 and SAP. But they are not overtly AU rich as TNFα might be, so we are considering other possibilities. We don't think it is going to be a simple consensus sequence solution.

Wakeland: The human Sanroque allele you found is very nice, because this is clearly a disease allele. Did you find homozygotes in these patients?

Vinuesa: The patient looks homozygous, but we have had a few different sequences from the same patient that looked different, so we wonder whether there might be a somatic contribution to this particular mutation. This is one that we have only found in one patient.

Wakeland: So the other four were other mutations.

Vinuesa: Yes, and I didn't present the data because we haven't done anything on the function of these.

Wakeland: Are they homozygous or not?

Vinuesa: The second mutation that is found in four different patients is a heterozygous mutation. We have found zero of these heterozygotes in controls and over 100 other patients that we have looked at of similar ethnic origins.

Wakeland: Have you looked at other family members, to examine the inheritance of this pattern?

Vinuesa: We are currently doing this.

Wakeland: We also see a follicular-like gene expression profile in our mice as they develop disease. We don't see it early, prior to disease. One of the issues we have been concerned about is that if we wait until the mice have ongoing autoimmune disease we see a lot of changes in gene expression profile that are consistent with the development of this chronic immune activation. We have focused looking at young mice to try to understand the underlying mutational lesion that mediates the development of autoimmunity. Have you looked at Sanroque mice prior to disease phenotype, to see whether they have these same phenotypes?

Vinuesa: I guess we haven't looked early enough. The germinal centre phenotype is found even in mice that aren't weaned. One of the key issues for us is that when we do the bone marrow chimeras we find some of these phenotypes, including T cell activation and ICOS overexpression, only in the mutant component. This tells us that it is intrinsic and it will be a target of Roquin. I don't think it is only age. We know there are some features of disease that are probably a late consequence of mice having developed hyperactivation, but the ones we are more excited about, such as ICOS overexpression, are likely to be primary events.

Wakeland: We don't see anything like this in our mice; certainly not to that level. This seems quite distinct to the Sanroque phenotype. What happens in the Sanroque heterozygote? Does it have a phenotype?

Vinuesa: The female heterozygotes develop autoantibodies relatively early (8–12 weeks), but not as early as the homozygotes (6 weeks). Heterozygotes and homozygotes have almost the same incidence of nephritis, but in Sanroque mice this occurs relatively later in life. Heterozygotes develop type 1 diabetes with an incidence of something like 50%.

Wakeland: Is this a dominant negative, then?

Vinuesa: We think there is a gene dosage effect with age. If you look early on it appears to be purely autosomal recessive.

Wakeland: Autoimmunity tends to be like this. SLE1 as a heterozygote tends to delay the development of autoantibodies, but doesn't lead to a complete loss of the phenotype. Allele dose seems to be involved in our genes as well.

Vinuesa: We have some evidence from our *Drosophila* collaborator, Rob Saint, that the total loss-of-function allele is lethal in *Drosophila*. We think we are dealing with a hypomorph. In humans we think that heterozygote mutations might confer a phenotype, simply because of this.

Wakeland: When you take Sanroque on the C3H background the phenotype seems to be fairly preserved.

Vinuesa: Yes. There are some changes, for example the thrombocytopaenia is much more obvious in the CBA background. In fact, if we bleed mice, 50% tend to die about a week later. In terms of the lymphoadenopathy and autoantibodies, we see this phenotype comparably in B6 and CBA.

Wakeland: Have you looked at any other backgrounds, and are there any that seem to alter the Sanroque phenotype?

Vinuesa: Originally we tried to map NOD, and gave up after a few generations because we thought we saw phenocopies. We have now concluded that they weren't phenocopies, but that a single Roquin allele in the NOD genetic background is probably giving the full-blown lupus disease. We know that NODs have the Idd5.1 ICOS form so they perhaps express more ICOS.

Cyster: What ideas do you have on specificity of your autoantibodies, and whether there are other autoantibodies? With the C1Q-deficient mouse, there is at least a model for why it ends up targeting nuclear or cytoplasmic autoantigens. What are your thoughts in this case?

Vinuesa: Sanroque mice have the full spectrum of ANAs. We have done some Western blots which have confirmed there are extractable nuclear antigens. An attractive possibility is that since most ANAs are exposed on the surface of the cell during apoptosis, if there is a strong germinal centre component, the germinal centre is one of the locations where there are the highest rates of apoptosis. Lots of these ANAs will be exposed. This could be facilitating the generation of multiple specificities.

Cyster: This would be true in any germinal centre reactions. Do you think there is a defect here in apoptotic clearance?

Vinuesa: No. There are likely other pathways to lupus that are not exclusively germinal centre driven.

Cook: Since the C1q deficiency is involved in clearance of apopotic cells, a defect in C1q results in is an increased amount of self-antigen, probably in an immunogenic form. Of course, there is also evidence that the spectrum of self-antigens that become available when cells undergo apoptosis appears to be enriched for those antigens that are relevant to lupus, such as chromatin and phospholipids. Here the evidence so far is that the spectrum of antibodies that appear in the presence of the Sanroque defect is similar to that observed in lupus. There are not only ANAs but also antibodies to dsDNA. We are not necessarily invoking a defect in complement to explain this—rather, the defect in tolerance within germinal centres may be sufficient explanation, since germinal centres contain so many apoptotic cells.

Cyster: I am wondering why it would be restricted to those. You suggest the germinal centre as the rationale. It is really the lack of access to other tissue-specific autoantigens for which we know B cells are not tolerant to begin with.

Vinuesa: It could be, yes. If we put these on a TCR Hel transgenic background, with age they develop arthritis.

Cyster: So it may not be as specific as the C1Q situation from the point of view of autoantibodies. Do you think that the diabetes system will involve ICOS?

Vinuesa: It isn't clear at this point.

Gros: You say that in several of the processes, you had between 15–20% of the pedigrees coming out positive for activity against nuclear antibodies. Didn't you find this to be an unexpectedly large number?

Vinuesa: Yes, very scary!

Gros: If you cross to CBA then they all go away. So is it possible that the background you started with had something to do with this? It looks like a sensitized screen to me, in that you already have a hit and then there is a threshold effect where you can hit in many different places and get the phenotype.

Vinuesa: Obviously, we always include controls from the same unmutagenized background strain. When we do this we don't see the same ANA incidence. When we breed them for a couple of generations to test habitability we do find they are heritable. I am now confident that what we are seeing is genuine. Why are we seeing so many? When we do the calculations there might be 100–200 genes. But in the literature, there are 40 mouse strains of knockout mice that develop ANAs. This doesn't mean that the mice have lupus or any other feature of autoimmunity; they just have ANAs. Our favoured interpretation is that there is probably some positive selection for all these genes that are important for protecting against infectious disease, but small tweaking of signalling thresholds in T and B cells can very easily tip the balance towards losing a bit of self-tolerance and producing ANAs.

Gros: Then you would have systematic suppression of all these things by crossing out to another strain.

Vinuesa: Possibly, yes. Ward Wakeland's work throws a lot of light onto the amount of very strong suppressors that you can find in different strains. We think that CBA may be one of those that is very rich in modifiers.

Gros: Of the 20% that are positive for ANA, what percentage come out positive for the other phenotype you have in your screen, which I thought was IgG and lymphocyte activation?

Vinuesa: IgG is pretty non-specific. B and T cell activation phenotypes are relatively rarer. Perhaps one in 10 strains we have looked at has a clear FACS phenotype. We have decided not to go into detailed characterization until we have evidence of some kind of linkage, because up to now we have had lots of trouble mapping, simply because of this phenotype disappearing.

Gros: When you use a mixed CBA/B6 F1 to screen, what percentage of ANA positivity did you find?

Vinuesa: Surprisingly, the same kind of prevalence—about 19%.

Gros: That's already crossed with CBA, right?

Vinuesa: Yes. This tells us that when we screen we will hit the same kind of genes. We are identifying particular kinds of mutations that will not be suppressed. This is at a level where we can map directly, and we have already found a few linkages. We think this is genuine linkage.

Goodnow: It is not all that are suppressed. As you said, if anything Sanroque is slightly exaggerated on the CBA background. It might well be that there is another set of partially overlapping 200 genes that you only reveal if you do the same screen on a different sensitized background. You can imagine that the list might grow up from 200 to eventually, in an outbreed population such as us, over 1000.

Vinuesa: I wouldn't think that is so surprising. In your original analysis when you were just looking at genes that contributed to T cell lymphopaenia you already came up with a list of 100 genes that in themselves could lead to T cell lymphopaenia. This is one of the causes of autoimminity. If to this one pathway in the thymus, you start adding T cell activation thresholds, B cell defects and macrophage and apoptotic cell clearance defects, it is not so hard to come up with a huge list of genes.

Wakeland: In terms of looking at the frequency of ANA and lupus-like phenotypes in knockouts, it is pretty important to keep in mind that when you cross 129 and B6, you get a significant autoimmunity even though neither parental strain has any inherent autoimmunity. This illustrates that epistasis between different backgrounds can lead to ANA production without any specific mutation involved. Recently, a BXD recombinant inbred strain was reported to develop a severe autoimmunity with a lupus-like phenotype although neither parental strain has that phenotype. When we simply start mixing strain genomes, a subset of the mice will be ANA positive. One more thing to remember is that when we talk about ANA in human populations, if we just consider a titre of 1–40 (which is a significant amount of IgG), the frequency is about 20%. Low titre ANA is a very common phenotype in human populations. When people are said to be ANA-positive in the context of a lupus predisposition, it is when they have a high titre. In your assays are you looking at low or high titre ANA?

Vinuesa: We started looking at what other people had been describing, which was titres of ANA found in 1 in 40 in mice. We then moved to 1 in 100. With this we only detect ANAs in LPR mice. I think at this titre we are confident that we are finding what is equivalent to a little bit of a higher titre in humans. Your point about the backgrounds is valid. We are now trying to detect them on a B6-related background. Chris has introduced the C57 Leaden strain as the mapping strain. It is closely related to B6, but still sufficiently polymorphic to be able to map. We therefore think that we won't be dealing with a lot of epistatic interactions and modifiers. Whatever we identify should be the real thing.

Gros: I still think this number is very high. There could be non-genetic explanations for this. You could have a retrovirus popping up that has a tropism for a particular population of cells that will cause a given phenotype.

Vinuesa: Would this be heritable?

Gros: Absolutely. It won't segregate, but it will be completely suppressed if you cross it with another strain that is non-permissive for this retrovirus. Then by crossing two strains that are permissive, you may bring two replication-defective copies that will rearrange and create expansion of a particular cell type. At least in one BXH strain that we have studied (Turcotte et al 2004, 2005), this is exactly what happens. For a long time we have tried to map this trait and were unable to because it wasn't inherited, but phenotypically it was pretty obvious.

Hume: This actually happens in the sense that one of the outputs of the RIKEN project which we haven't come to grips with is that endogenous retroviruses are also tissue-specifically transcribed. They are also strain specific. Some of the genes that we find are inducible only in B6 or DBA are in fact endogenous retroviruses.

Goodnow: So just breeding by chance would have activated the retrovirus? How might this work?

Gros: You could have two strains that carry replication-incompetent retroviruses. If you then cross them together and have a predisposing mutation on a sensitized background this virus will expand, in our case in the myeloid lineage, creating a myeloproliferation which if you were to score would look like a myeloproliferative disease. You can't map this trait because it is inherited as a recessive trait yet it creates a phenotype that can interfere with your screen.

Goodnow: When it is hopping, presumably it is insertionally activating oncogenes.

Gros: Yes, and then you can have a variety of predisposing mutations that will facilitate this. You could have a variety of primary hits that are inherited and then the retrovirus provides the second hit when it starts replicating in that population. This can reveal oncogenes in different places. Neil Copeland has used BXH2 to identify a great number of retroviral insertion sites that inactivate tumour suppressors or activate oncogenes in clonal tumours observed in these mice (Li et al 1999).

Wakeland: Isn't the frequency low? It can't be used as a mutagenesis strategy. But this doesn't mean that it wouldn't occur at a reasonable frequency if we are talking about 100 gene targets.

Vinuesa: Can you screen for that? We still don't have explanations of why we can't map certain lines.

Gros: You can take DNA from a particular cell population that you think is affected and screen with viral probes. You can look for the endogenous versus the extra copies that are replicated.

Goodnow: This relates to another issue, which I have wondered about, particularly because of the work that Rioux-Laucat has done at Hopital Necker showing that it is possible to have a somatic mutation in a haemopoietic cell in the Fas gene. This cell then has a selective advantage and then grows out as an autoimmune driving clone. I wonder whether this might be a much bigger issue than we think. One explanation for semi-dominant behaviour for loss of function alleles could be that you inherit one defective copy of a tolerance gene and most of your clones will be regulated fine. But sooner or later one of your T cells or B cells will lose the remaining functional copy and is well along the way to autoimmunity.

Wakeland: That is the thing about C1Q. Even though it is a hugely penetrant single-gene Mendelian-like phenotype for lupus, there are only 40 or so individuals worldwide that have been identified as being homozygous for that phenotype. On the other hand, if there is an allele dose effect like with Sanroque it could easily play a more significant role. Although 1q23 does come up, it doesn't come up that strongly in many of the linkage studies that have been done. Many of the data argue against any strong Mendelian inheritance by a particular locus.

Vinuesa: It is possible that current linkage studies do not identify loci with Mendelian inheritance due to the phenotypic heterogeneity of a disease like lupus, in which any individual patient needs only fulfil 4 of 11 criteria to be diagnosed. In these cases, stratification according to particular sub-phenotypes or clinical manifestations, i.e. having thrombocytopaenia, may aid in identifying rare genes with strong effects.

Wakeland: Whenever you start stratifying by phenotype, you are adding additional statistical factors into how you do these linkage studies. It is hotly debated as to whether this is a valid approach.

Vinuesa: I agree with what you are saying but lupus is so complicated and there are so many different manifestations that there have to be different pathways leading to it. It is already evident from mouse models that there are T-dependent and T-independent pathways.

Goodnow: We have so many cases now where if you do a population-based approach, lumping all patients together and treating them equally, you don't see anything in a genome scan. But when you look at specific candidate genes and do a patient-oriented approach you do see something. With Jean-Laurent Casanova's IFNγ pathway genes, if he had lumped all these individuals together his genome scans would probably have found nothing.

Casanova: We can probably go back earlier in time. The first human germline genetic lesion found, trisomia 21 in patients with Down syndrome, was not thought to be genetic until Jerôme Lejeune did a karyotype and found a third chromosome 21. That is quite a lesion! It is sporadic, common and it is an overt chromosomal defect. It was the first human congenital disease to be understood at the molecular genetic level.

Hume: It is not transmissible, so it's not genetic, strictly speaking.

Wakeland: The nice thing about the Sanroque study in human is that you could demonstrate that at least one of these alleles does have a phenotype in mice quite consistent with what is seen in the original mutation. Is there any way you could do a similar thing with your alleles? One of the problems of these rare variants is that you end up with one or two families. The power is simply insufficient to map into a tight critical region. In the end, you find this gene, characterize the mutation, and make the case for the idea that it causes this disease, but it can be difficult to prove this to a statistical geneticist.

Goodnow: Warwick Britton showed us very nicely a set of variants in human that clearly have a semidominant functional defect in about as good a functional assay as you can get for a clinical readout—killing of TB organisms *in vitro*, and yet we still have a doubt that we are putting A and B together when it is just population stratification. This will be a conceptual challenge for us. This always reminds me of Terry Speed who spends half his time at the Walter and Eliza Hall Institute in Melbourne as a mathematical geneticist: he has a T shirt saying 'statistics means never having to say you are sure'! We may never be sure.

References

Li J, Shen H, Himmel KL et al 1999 Leukaemia disease genes: large-scale cloning and pathway predictions. Nat Genet 23:348–353

Turcotte K, Gauthier S, Mitsos LM et al 2004 Genetic control of myeloproliferation in BXH-2 mice. Blood 103:2343–2350

Turcotte K, Gauthier S, Tuite A, Mullick A, Malo D, Gros P 2005 A mutation in the Icsbp/IRF-8 gene causes susceptibility to infections and a CML-like syndrome in BXH-2 mice. J Exp Med 201:881–890

Genetic resistance to smallpox: lessons from mousepox

Gunasegaran Karupiah, Vijay Panchanathan, Isaac G. Sakala and Geeta Chaudhri

Infection and Immunity Group, Division of Immunology and Genetics, John Curtin School of Medical Research, Australian National University, PO Box 334, Canberra, ACT 2601, Australia

Abstract. There is increased interest in understanding protective immunity to smallpox for two principal reasons. First, it is the only disease that has been successfully eradicated using a live virus vaccine and, second, there exists a potential threat of intentional or unintentional release of variola virus, the causative agent of smallpox. Although mortality rates associated with smallpox were as high as 40%, a significant subset of those infected recovered. The basis of susceptibility or resistance, and the immune parameters associated with recovery, are still unknown. Animal models of poxvirus infections are being employed to understand what constitutes an effective host response. Ectromelia virus is closely related to variola virus and it causes a disease similar to smallpox in mice. This model is well established, resistant and susceptible strains of mice are defined and four genetic loci associated with resistance have been identified. Susceptibility to infection and disease severity is also influenced by virus immune evasion strategies. The outcome of infection is clearly dictated by several factors including host and viral genes, both of which influence the immune response. Here we present data on one virus-encoded immune modifier and its effect on the functions of two host genetic loci associated with resistance.

2007 Decoding the genomic control of immune reactions. Wiley, Chichester (Novartis Foundation Symposium 281) p 129–140

Host response to viral infection and determinants of disease outcome

The host response to viral infection may be divided into two distinct but complementary and interactive parts. The first comprises a series of rapidly acting antimicrobial mechanisms, including natural killer (NK) cells, which form the innate phase of the immune response. These mechanisms protect the host in the earliest stages of infection. They prevent most infections from becoming established and allow time for generation of the slower adaptive immune response, including cytotoxic T lymphocytes (CTLs), that are antigen-specific and which establish immunological memory. The innate response which is comprised of a range of soluble factors and various leukocyte subsets can profoundly influence and direct the type of adaptive response that is generated and is therefore critical in determining the outcome of a viral infection. Disease severity and outcome are influenced by factors such as virus strain, virus immune evasion strategies, dose and route of infection. The outcome of infection is therefore dictated by several factors including host and viral genes, both of which profoundly influence the immune response.

Protective immunity to smallpox

Smallpox was one of the biggest human scourges, resulting in mortality rates of up to 40% in some populations. However, a significant subset of the infected population recovered. The basis of susceptibility and resistance, and the immune parameters associated with recovery, is not known as the virus was eradicated more than 25 years ago. Despite the success of the smallpox eradication program, there remains considerable fear that variola virus (VARV), the causative agent of smallpox, or other related pathogenic poxviruses such as monkeypox (MPXV) could re-emerge and spread disease in the human population. The increased interest in understanding protective immunity to smallpox is due not only because of the potential threat of a bioterrorist attack (Henderson et al 1999) but it is the only disease known to humankind that has been successfully eradicated with a live virus vaccine. As such, very useful information on immunity and resistance to disease may be gleaned from the study of smallpox.

Smallpox, mousepox and genetic resistance

Since VARV has a restricted host range and is known to infect only humans, closely related orthopoxviruses, such as MPXV and ectromelia virus (ECTV), have been used extensively in animal models to elucidate pathogenesis and immune response to infection. Currently, the best surrogate for VARV in a small animal model is ECTV, as it is infectious at very low doses, has a restricted host range, encodes a similar repertoire of immune evasion proteins and causes severe disease (mousepox) with high mortality rates (Esteban & Buller 2005, Fenner et al 1988, Seet et al 2003). Further similarities between smallpox and mousepox include virus replication and transmission, cytokine responses and many aspects of pathology (Esteban & Buller 2005). Thus, to understand the genetic basis of resistance and susceptibility to smallpox in humans, we have used mousepox, where resistant and susceptible mouse strains are well defined and at least four genetic loci associated with resistance have been mapped.

Mousepox: a model for study of virus–host interactions and smallpox

ECTV is a natural mouse pathogen that has co-evolved with its host. Inbred strains of mice are resistant or susceptible to infection with ECTV. The virus causes an acute disease characterized by generalized viral spread and pathology. Death usually occurs due to extensive necrosis of major organs, in particular the liver and spleen, as a result of massive virus replication (Fenner 1948). Both MHC and non-MHC genes determine resistance or susceptibility to mousepox. Strains such as A/J (H-2a), BALB/c (H-2d) and DBA/2 (H-2d) exhibit high mortality (100%) to mousepox while C57BL/6 (H-2b), C57BL/10 (H-2b) and 129 (H-2b) strains have very low mortality, limited pathology and are classified resistant. Wild mice show variable susceptibility to mousepox (Buller et al 1986). The mousepox model has been utilized extensively to study virus–host interactions, genetic resistance to disease and viral immunology (O'Neill et al 1983, Buller et al 1986, Brownstein et al 1992, Karupiah et al 1993, 1996, 1998, Brownstein & Gras 1997, Mullbacher 2003, Chaudhri et al 2004, 2006, Panchanathan et al 2005, 2006, Tscharke et al 2005). Since mousepox is a model of a *natural infection*, it is a powerful tool for understanding the immune response. It not only provides a useful model for smallpox but also for other generalized viral infections.

Immunity to ECTV infection: Roles of innate and adaptive immune responses

The resistant strains of mice, such as C57BL/6, generate a strong T helper 1 (Th1)-type cytokine response (i.e. interleukin [IL]2, γ interferon [IFNγ] and IL12) and potent NK cell and antiviral CTL responses whereas these responses are either delayed and weak or lacking in susceptible strains of mice (Chaudhri et al 2004). We have defined key immune parameters in C57BL/6 mice that are important for recovery from mousepox. These include the effector functions of NK cells, $CD4^+$ and $CD8^+$ T lymphocytes, B lymphocytes and antibody, macrophage subsets, nitric oxide and IFNα, β and γ (Chaudhri et al 2004, 2006, Karupiah et al 1996, 1998, 1993, Panchanathan et al 2005, 2006, Ramshaw et al 1997). Perforin, granzymes and IFNγ are necessary for the effector functions of $CD8^+$ T cells and NK cells in the control of this virus (Karupiah et al 1993, Mullbacher 2003, Panchanathan et al 2006, Ramshaw et al 1997).

The role of antibody in recovery from a primary ECTV infection was only recently established (Chaudhri et al 2006, Fang & Sigal 2005). We have previously reported that in C57BL/6 mice, which are normally resistant to mousepox, the absence of $CD4^+$ T cells resulted in ECTV persistence for extended periods (Karupiah et al 1996) and the animals eventually succumb to disease. The antiviral CTL response in mice lacking $CD4^+$ T cells was suboptimal, suggesting that virus persistence maybe the result of a defective CTL response. However, in contrast to the C57BL/6 wild-type mice, antiviral CTL activity in these mice persisted even in the late stages of infection (Karupiah et al 1996). Notwithstanding, the CTL were insufficient to clear virus. Since $CD4^+$ T cell help is also crucial for antibody production (MacLennan et al 1997, Parker 1993), we hypothesized that virus

TABLE 1 Genetic loci that control resistance to mousepox in C57BL/6 mice

Designation	*Location*	*Locus*	*Reference*
Rmp1	Chromosome 6	Natural killer cell complex	Brownstein & Gras (1997), Delano & Brownstein (1995)
Rmp2	Chromosome 2	Complement component C5	Brownstein & Gras (1997)
Rmp3	Chromosome 17	MHC (H-2) complex	Brownstein et al (1992)
Rmp4	Chromosome 1	Selectin gene complex	Brownstein & Gras (1997)

persistence in these animals might be due to defective antibody response. Indeed, mice deficient in B cells (Kitamura et al 1991) succumbed to mousepox between 3–4 weeks post-infection although they appeared to keep the infection under check during the first two weeks. In many respects, the kinetics of virus replication and outcome of infection in B cell deficient mice was similar to mice deficient in $CD4^+$ T cells (Chaudhri et al 2006). Our data show that mice deficient in $CD8^+$ T cells or $CD8^+$ effector function die early in infection whereas those deficient in B cells or antibody production die much later, indicating that B cell function becomes critical after the effector phase of the $CD8^+$ T cell response to infection subsides. In mice lacking B cells or antibody, ECTV persists and the host succumbs to disease, despite the generation of normal $CD8^+$ T cell responses (Chaudhri et al 2006). The importance of antibody in a primary infection had not been previously appreciated.

Genetic resistance to mousepox

In mice, there are at least four loci known to confer resistance against mousepox (Table 1). The resistance to mousepox (*Rmp1*) locus on chromosome 6 maps to the NK gene complex (NKC) (Brownstein & Gras 1997, Delano & Brownstein 1995). The *Rmp2* locus on chromosome 2 maps near the complement component C5 gene (Brownstein & Gras 1997). *Rmp3* is also gonad-dependent and is linked to the MHC (H-2) on chromosome 17 (Brownstein et al 1992). Finally, the *Rmp4* locus, on chromosome 1, maps to the selectin gene complex (Brownstein & Gras 1997). Only *Rmp1* and *Rmp3* will be briefly discussed here.

The Rmp1 locus

The critical role NK cells play in the control of infection with viruses is well-recognized (French & Yokoyama 2003). In the murine cytomegalovirus (MCMV) model, susceptibility to lethal infection can be overcome by a dominant allele at the *Cmv1* locus in the C57BL/6 mouse. *Cmv1* maps to the NKC on chromosome 6 and controls early replication of MCMV via regulation of NK cells. It is now established that Cmv1 is Ly49H, an NK cell-activating receptor (Arase et al 2002, Forbes et al 1997). The NKC in the BALB/c mouse strain lacks some activating receptors, including Ly49H and is therefore susceptible to MCMV whereas the C57BL/6 strain is resistant. The generation of BALB/c mice congenic for C57BL/6 NKC (designated BALB/c.B6.Cmv1^r) (Scalzo et al 1999) has been important to definitively establish that Cmv1^r is Ly49H. More recently, we have used the BALB/c.B6.Cmv1^r mice in the ECTV model to study the role of the NKC in conferring genetic resistance to mousepox.

The *Rmp1* locus on chromosome 6 controls replication of virus in the spleen and liver. It maps to the NKC (Brownstein & Gras 1997, Delano & Brownstein 1995) and this is consistent with the known importance of NK cells protection against infection (Brownstein & Gras 1997, Delano & Brownstein 1995, Karupiah et al 1996). We have recently shown BALB/c.B6.Cmv1^r mice displayed increased resistance to ECTV infection compared with wild-type BALB/c mice, which showed uniform mortality. The increased resistance of BALB/c.B6.Cmv1^r mice was completely abrogated when NK1.1 cells were depleted, indicating that the effect was due to NK cells. In addition, we have preliminary evidence that treatment of ECTV-infected BALB/c.B6.Cmv1^r mice with monoclonal antibodies to specific Ly49 family of proteins significantly increased viral load in the spleen and liver but not as much as titres in mice depleted of NK1.1 cells. We predict that the activity of *Rmp-1* may be attributed to at least two or more NK cell-activating receptors.

The Rmp3 locus

Rmp3 is linked to the *H-2* complex (Brownstein et al 1992) and its function is expressed through CD8 T cells, which are essential for recovery from primary ECTV infection (Karupiah et al 1996, O'Neill et al 1983). The resistant strains of mice, such as C57BL/6, generate a strong Th1 cytokine response and potent antiviral CTL responses early in infection whereas in susceptible strains of mice such as A/J and BALB/c, these responses are delayed and weak or completely lacking (Chaudhri et al 2004). In general, mice of the H-2^b haplotype (C57BL/6, C57BL/10 and 129) are resistant, mice of the H-2^d/H-2^a haplotype (BALB/c, DBA/2, A/J) are highly susceptible and those of the H-2^k haplotype (C3H, CBA) tend to be intermediate in susceptibility. We have preliminary evidence that C57BL/6 mice lacking MHC class I D^b are highly susceptible to mousepox, suggesting that the function of *Rmp3* may be attributed to, or expressed by, D^b. Although the immunodominant CD8$^+$ T cell determinant to ECTV is restricted by K^b (Tscharke et al 2005), mice deficient in this class I molecule were able to overcome infection with ECTV unlike the D^b GKO mice. Mice lacking both K^b and D^b molecules were highly susceptible to mousepox.

Influence of poxvirus-encoded host response modifiers to disease resistance

Viruses have evolved numerous mechanisms to evade detection and destruction by the host immune system. Orthopoxviruses have devised novel strategies, which directly target molecules of the immune system and can therefore influence the activities of NK cells and CD8 cells. ECTV, like VARV, encodes molecules that target the host apoptotic pathway, complement cascade, semaphorins, chemokines and cytokines. Of the viral proteins that target host cytokines, those which influence NK and $CD8^+$ T cell responses are viral IFNγ binding protein (vIFNγbp), vIFNα/βbp and vIL18bp. Indeed, susceptible BALB/c mice infected with a deletion mutant ECTV lacking vIFNγbp (ECTV-IFNγ$bp^{-/-}$) were able to effectively clear virus and recovered, unlike mice infected with wild-type ECTV, which succumbed to mousepox. The absence of vIFNγbp resulted in increased host IFNγ production that allowed the BALB/c mice to generate elevated NK cell and antiviral $CD8^+$ T cell responses (unpublished data). These data are consistent with the idea that the full expression of genetic resistance by the host to virus infection is clearly affected by virus-encoded immunomodulatory genes.

Summary

We have used the mousepox model as a model for smallpox to understand the roles and mechanisms of some host and viral genes in conferring resistance to disease and show that both viral and host genes profoundly influence the outcome of infection.

Acknowledgements

This work was supported by grants from the Howard Hughes Medical Institute and National Health and Medical Research Council of Australia.

References

Arase H, Mocarski ES, Campbell AE, Hill AB, Lanier LL 2002 Direct recognition of cytomegalovirus by activating and inhibitory NK cell receptors. Science 296:1323–1326

Brownstein DG, Bhatt PN, Gras L, Budris T 1992 Serial backcross analysis of genetic resistance to mousepox, using marker loci for Rmp-2 and Rmp-3. J Virol 66:7073–7079

Brownstein DG, Gras L 1997 Differential pathogenesis of lethal mousepox in congenic DBA/2 mice implicates natural killer cell receptor NKR-P1 in necrotizing hepatitis and the fifth component of complement in recruitment of circulating leukocytes to spleen. Am J Pathol 150:1407–1420

Buller RM, Potter M, Wallace GD 1986 Variable resistance to ectromelia (mousepox) virus among genera of Mus. Curr Top Microbiol Immunol 127:319–322

Chaudhri G, Panchanathan V, Buller RM et al 2004 Polarized type 1 cytokine response and cell-mediated immunity determine genetic resistance to mousepox. Proc Natl Acad Sci USA 101:9057–9062

Chaudhri G, Panchanathan V, Bluethmann H, Karupiah G 2006 Obligatory requirement for antibody in recovery from a primary poxvirus infection. J Virol 80:6339–6344

Delano ML, Brownstein DG 1995 Innate resistance to lethal mousepox is genetically linked to the NK gene complex on chromosome 6 and correlates with early restriction of virus replication by cells with an NK phenotype. J Virol 69:5875–5877

Esteban DJ, Buller RM 2005 Ectromelia virus: the causative agent of mousepox. J Gen Virol 86:2645–2659

Fang M, Sigal LJ 2005 Antibodies and CD8+ T cells are complementary and essential for natural resistance to a highly lethal cytopathic virus. J Immunol 175:6829–6836

Fenner F 1948 The pathogenesis of the acute exanthems: an interpretation based on experimental investigations with mousepox (infectious ectromelia of mice). The Lancet 252:915–920

Fenner F, Henderson DA, Arita I, Jezek A, Ladnyi ID 1988 Smallpox and its eradication. World Health Organization, Geneva

Forbes CA, Brown MG, Cho R et al 1997 The Cmv1 host resistance locus is closely linked to the Ly49 multigene family within the natural killer cell gene complex on mouse chromosome 6. Genomics 41:406–413

French AR, Yokoyama WM 2003 Natural killer cells and viral infections. Curr Opin Immunol 15:45–51

Henderson DA, Inglesby TV, Bartlett JG et al 1999 Smallpox as a biological weapon: medical and public health management. Working Group on Civilian Biodefense. JAMA 281:2127–2137

Karupiah G, Fredrickson T, Holmes K, Khairallah L, Buller R 1993 Importance of interferons in recovery from mousepox. J Virol 67:4214–4226

Karupiah G, Buller RM, Van Rooijen N, Duarte CJ, Chen J 1996 Different roles for CD4+ and CD8+ T lymphocytes and macrophage subsets in the control of a generalized virus infection. J Virol 70:8301–8309

Karupiah G, Chen JH, Nathan CF, Mahalingam S, MacMicking JD 1998 Identification of nitric oxide synthase 2 as an innate resistance locus against ectromelia virus infection. J Virol 72:7703–7706

Kitamura D, Roes J, Kuhn R, Rajewsky K 1991 A B cell-deficient mouse by targeted disruption of the membrane exon of the immunoglobulin mu chain gene. Nature 350:423–426

MacLennan IC, Gulbranson-Judge A, Toellner KM et al 1997 The changing preference of T and B cells for partners as T-dependent antibody responses develop. Immunol Rev 156:53–66

Mullbacher A 2003 Cell-mediated cytotoxicity in recovery from poxvirus infections. Rev Med Virol 13:223–232

O'Neill HC, Blanden RV, O'Neill TJ 1983 H-2-linked control of resistance to ectromelia virus infection in B10 congenic mice. Immunogenetics 18:255–265

Panchanathan V, Chaudhri G, Karupiah G 2005 Interferon function is not required for recovery from a secondary poxvirus infection. Proc Natl Acad Sci USA 102:12921–12926

Panchanathan V, Chaudhri G, Karupiah G 2006 Protective immunity against secondary poxvirus infection is dependent on antibody but not on CD4 or CD8 T-cell function. J Virol 80:6333–6338

Parker DC 1993 T cell-dependent B cell activation. Annu Rev Immunol 11:331–360

Ramshaw IA, Ramsay AJ, Karupiah G et al 1997 Cytokines and immunity to viral infections. Immunol Rev 159:119–135

Scalzo AA, Brown MG, Chu DT et al 1999 Development of intra-natural killer complex (NKC) recombinant and congenic mouse strains for mapping and functional analysis of NK cell regulatory loci. Immunogenetics 49:238–241

Seet BT, Johnston JB, Brunetti CR et al 2003 Poxviruses and immune evasion. Annu Rev Immunol 21:377–423

Tscharke DC, Karupiah G, Zhou J et al 2005 Identification of poxvirus CD8+ T cell determinants to enable rational design and characterization of smallpox vaccines. J Exp Med 201:95–104

DISCUSSION

Casanova: What is your strategy for identifying the NK susceptibility gene?

Karupiah: We're going to use the same strategy that Tony Scalzo and Wayne Yokoyama used with the mouse cytomegalavirus model, which is to use antibodies specific for the Ly49 family of molecules. We have done some work with Ly49 H and D. We think it is one of them, but we think there is more that one: it is not a single gene.

Casanova: Do you think it could be a susceptibility haplotype with several genes nearby involved?

Karupiah: We have some evidence that it could be both Ly49 H and D.

Foote: It seems that B6 mice are particularly resistant to many infectious diseases, such as malaria and leishmaniasis. Everyone seems to take the same approach to this, looking at cytokine profiles. Inevitably, they reach a place where they find that cytokine profiles and many of the immunological phenotypes they are measuring are almost irrelevant to the phenotype that is being looked at. You have a couple of instances where you find exceptions to this rule, but by and large there seems to be a discord between cytokine profiles and differences in susceptibility.

Karupiah: Taking just one cytokine as an example, we know that IFNγ is critical in the three different strains we have looked at. The clue to why it is important perhaps comes from the viral-encoded IFNγ binding protein. I agree with you that some of the cytokines that we think are important might not matter, but there are some that are critical.

Foote: There are congenics available, especially from the chick group where they have the congenics for some of the cytokine responses. It would be interesting to challenge these to see whether the cytokine changes are important at the functional level. This is different to knocking the gene out.

Hume: Along the same lines, it is self-evident that an appropriate T cell activation response is required. In looking at the cytokine profiles, you are studying what Peter Doherty called the interface of microbiology, immunology and pathology, which is mythology. You are downstream of the consequences of the initial recognition of the pathogen. In the early work on bacterial susceptibility loci, on this same sort of cross, you could tell that the outcome of the infection was determined in the first 30 min, depending on whether BCG was localized to the initial site of injection or disseminated, for example. Is there a cell autonomous difference in the primary recognition and response to the pathogen? In other words, do macrophages take it up to the same extent and induce the same early response genes? If those are different, everything else follows.

Karupiah: The only thing we know about the macrophage difference seems, at least *in vivo*, to be at the level of the Kupffer cells in the liver. One level of resistance seems to be whether Kupffer cells in the liver can support replication. We know that this happens in susceptible strains.

Hume: If that were to be the case, wouldn't it make sense to gene profile the precise phenotype of Kupffer cells between the resistant and susceptible mouse strains. What is the primary difference in determining the outcome of infection? At what stage in the infection process can you tell that the animal is going to fail to control the replication of the virus?

Karupiah: It is not as simple as whether the macrophage gets infected or not. Many responses are generated which activate the macrophage and make the activated macrophages in a resistant mouse able to control the infection, preventing spread to the hepatocytes.

Hume: Do you have evidence that macrophages from the resistant and susceptible strains are infected in exactly the same way?

Karupiah: Yes, *in vitro* and *in vivo*. I don't think it is simply at the level of macrophages. It is more complicated. Of course, macrophages are critical. About 10 years ago we used liposomes to deplete macrophages. They are not able to generate a host of responses, including the cytotoxic T cell response. I believe that macrophages, in addition to neutrophils, are the earliest players that need to be involved in some way in presentation of antigen for the generation of the cytotoxic T cell response. There are differences in the susceptible and resistant strains in terms of activation and production of not only cytokines but also activation of other cell types needed in order to have this genetic resistance expressed. I don't think it is just simply macrophages.

Hume: One question is whether a B6 nude resists a virus any better than a BALB/c nude.

Karupiah: This doesn't speed up the death process. If you infect a nude mouse, whether it is BALB/c or B6, they die about the same time. But the process can be speeded up by using a nude which doesn't have IFNγ.

Maizels: Why don't you get sterilizing immunity from the CTL population? In the resistant mouse for two weeks they seem to keep everything in bay, but what happens after this? Do the CTL decrease in frequency, or are the interference proteins from the virus able to switch off the CTLs?

Karupiah: That's not the case. Even at 35 d the CTLs are still there, and they are still cytolytic. It appears that the antibody is needed to control virus in blood. Once the virus has reached the skin, each pock lesion has very high titres of virus. This seems to be a site from which virus then seeds back into the blood and gets to other organs.

Maizels: So the CTL response is never sufficient.

Turner: What do you think the role of the CTL is? Is it just a containment to make sure the viral load doesn't get too high to overcome the host?

Karupiah: Yes, I think so. In a secondary infection CTLs are not needed, just antibody. In the mousepox model we have done this with a whole range of knock-outs as well as depletion of leukocyte subsets. Another group has done this with the monkeypox virus model using monkeys.

Turner: Can you increase $CD8^+$ T cell numbers to a point where they can be protective and you don't get persistence of virus?

Karupiah: In the experiments where we transferred $CD8^+$ cells to B cell-deficient mice which themselves generate normal cytotoxic T cell responses, this didn't seem to help. The increase in numbers is of no use. The simple view of free virus requiring just antibody is wrong. No amount of cytotoxic T cells is going to clear that virus in the absence of antibody.

Turner: That is an important point you are making, in terms of the adaptive immune system. If the virus is circulating freely the antibody will be the thing that mops it up. Your model shows the cooperation that exists between the different arms of the immune response.

Wakeland: Was your mapping of these loci done in crosses with B6 mice?

Karupiah: We didn't do the mapping ourselves. This work was done by other groups who have since stopped doing the mapping.

Wakeland: I was specifically interested in the locus on chromosome 1, which you attributed to the selectin family. What is that based on?

Karupiah: This is based on genetic crosses using B6 and DBA2 mice.

Wakeland: Why the selectin family? This is on chromosome 1 and we have a locus there we are quite interested in, *Sle1*, which is near the selectin family. The Slam family is also adjacent, along a whole variety of genes that impact the immune system. How was the selecting family sorted out from all these, because this is a fairly daunting undertaking?

Karupiah: We haven't done this ourselves, so I can't say.

Wakeland: Also, if the *Sle1B* cluster, which is a Slam family member, is taken from DBA or 129 and put into B6, it makes these mice become more Th1 focused. This is consistent with the idea that this family is playing a role in resistance according to your model.

Karupiah: I want to add one point. Although we had these resistant and susceptible strains, if we took a highly avirulent strain of ectomelia virus and infected the susceptible strains, those animals can generate what looks like a normal cytotoxic T cell response. If the virus is slower in terms of replication, then these animals seem to be able to generate normal responses. They don't seem to be defective in their capacity to generate responses, but the virus seems to be overwhelming them.

Gros: What is the C5 status of BALB/c?

Karupiah: It is partial, that is, these mice still make C5 but not at similar levels to B6 mice.

Hume: BALB/c is still resistant to candida, I think.

Gros: BXD8 has the 23 kb deletion. Could you use that to figure out whether LY49H is the gene effect responsible?

Scalzo: That's one approach you could use. You could also use the knockout or transgenic mice. I have another question. Is there any evidence of genetic variation in the immune evasion genes of ectomelia virus?

Karupiah: That isn't known. We tend to use virus strains that are common to most labs. If you went and collected wild strains of the virus you might see this variation.

Scalzo: In the case of MCMV we find quite a lot of genetic variation in some immune evasion genes. This results in differences in the immune response and how the virus escapes from different arms of the innate and adaptive immune response.

Goodnow: It sounds like the challenge of decoding the genetic variation in the host and the virus is similar to the problem we faced in bacterial genetics with lysis versus lysogeny. It is a kinetic race between competing processes. Is it technically feasible in the first couple of days after infection to be able to see exhaustion of CTLs if the virus grows more rapidly, or are we dealing with decision points and kinetic races at a stage where we don't have the tools to measure T cell outcomes or viral replication? You made the point that one of the problems we face is this 'black box': through the kinds of approaches people have mentioned here we can show that this host gene affects it this way, or this effector mechanism works that way, but there is still this black box in the early days of infection. You can squeeze the black box ever earlier and see what some of the important inputs are, but to understand how all the inputs are being integrated may not be technically possible if a lot of the integration is happening in the first few days.

Karupiah: I suspect it is possible. We have tried to take a susceptible mouse and make it resistant. Some attempts, such as knocking out IL4, have been unsuccessful. We moved away from this and are looking at how, from the known resistance genes, we can change the outcome in the first few days.

Maizels: In terms of this race between the immune system and the virus, it sounds like the CTLs are too slow to kill the target cell. The virus is already packaged and therefore the lysis results in onward infection. Is that the case? Are they retarded in terms of how quickly they latch on to the infected cell?

Karupiah: We can measure the response by about 6 d post-infection. By this time it is determined whether the animal is going to die or live. Given the tools that we have, it may be possible to demonstrate that the responses come up a bit earlier. But then if we took this response away the titres would increase rapidly. In the absence of $CD8^+$ cells the animal dies in about 10–12 d, but if anything is defective in the innate response they are usually dead in 6 d. Yes, the CTL response seems to be a bit late, but without this the animal won't survive.

Turner: In the mice that lack $CD8^+$ function, what do the antibody responses look like? Are they lower?

Karupiah: The mice are usually dead by the time the antibody response comes up. If we use an avirulent virus to prime the animals and then challenge with virulent virus, the animals that lack $CD8^+$ T cells tend to make good antibody responses. Presumably this is a compensation.

Turner: In terms of the role for $CD8^+$ T cells, they do take a while to get going, but by the time we are able to measure them perhaps their effector function has already been utilised. Their role might be to keep virus loads down long enough so that the antibody response can kick in. In many models, $CD8^+$ models peak as the virus load is coming down.

The AcB/BcA recombinant congenic strains of mice: strategies for phenotype dissection, mapping and cloning of quantitative trait genes

Anny Fortin*, Eduardo Diez*, Janet E. Henderson†, Jeffrey S. Mogil†, Philippe Gros† and Emil Skamene‡

**Emerillon Therapeutics Inc., 416 de Maisonneuve West, Suite 1000, Montreal QC, Canada, H3A 1L2, †McGill University, Montreal, QC, Canada H3G 1Y6 and ‡The Research Institute of the McGill University Health Center, Room A6.149, 1650 Cedar Avenue, Montreal QC, Canada H3G 1A4*

Abstract. The AcB/BcA gene discovery platform consists of a series of 36 recombinant congenic strains (RCS) produced from the second backcross generation of the progenitor mouse strains A/J and C57BL/6J. Each individual inbred RCS carries 12.5% of the donor genome in 87.5% of the background genome. As the two parental strains are known to vary in the expression of resistance and susceptibility to a considerable number of mouse models of human diseases, the AcB/BcA RCS platform represents a valuable and versatile genetic tool to study many different phenotypes. RCS can be used to follow the segregation of single gene effects in individual strains, or to look at association/dissociation of mechanistic aspects of complex phenotypes. In addition, one can select strains with fixed alleles at known loci to look for novel gene effects, or use strains with overlapping congenic segments to delineate minimal QTL intervals. The AcB/BcA RCS platform was used by our group and others to study a series of complex phenotypes including nociception, malaria susceptibility and lipid metabolism. Linkage mapping in secondary crosses and gene expression analysis in targeted organs allowed the identification of chromosomal regions, genes, and biological pathways which might unravel novel targets for preventive and therapeutic interventions.

2007 Decoding the genomic control of immune reactions. Wiley, Chichester (Novartis Foundation Symposium 281) p 141–155

The mouse is the model organism of choice to dissect the genetic components controlling complex phenotypes. This model organism is widely used, mainly because (1) large numbers of wild-type isolates and mutant stocks of mice are available in an inbred status; (2) informative segregating animals can be generated in large numbers for linkage mapping and positional cloning; (3) the sequence of the mouse genome provides a compendium of candidate genes for a particular region; (4) null alleles at candidate genes can be readily obtained by gene targeting; and (5) mutant variants of the gene can be re-introduced on a null background to analyse genotype/phenotype correlations. Over the last few years, there has been a remarkable increase in the availability of new genomic strategies for the analysis of complex traits in mouse models, including screening of recombinant congenic (Demant & Hart 1986) and chromosome substitution strains (Singer et al 2004), advanced intercross lines (Darvasi & Soller 1995) and ENU-mutated animals (Balling 2001, Clark et al 2004). Combined to the advent of genome scanning, these new genetic tools have led to the identification of resistance/susceptibility loci for a large number of diseases relevant to human health.

Recombinant congenic strains (RCS) are produced by limited backcrossing between two inbred strains and subsequent inbreeding by brother–sister mating. In this way, a series of strains are created, each of which carries a small fraction of the genome of one strain (donor strain) on the genetic background of a second strain (background strain). The original hypothesis behind this breeding scheme was that unlinked genes contributing to a complex trait would become separated in different strains in which they would be studied individually (Demant & Hart 1986). In 1994, our group has undertaken the generation of a set of reciprocal RC strains from two of the best characterized and commonly used inbred strains of mice A/J and C57BL/6J (Fortin et al 2001b). These two parental strains were chosen because they differ in many important physiological traits, as well as in susceptibility to either naturally occurring or experimentally induced pathologies including metabolic, respiratory, and endocrine diseases, as well as in their response to several infectious and inflammatory stimuli (see *http://www.informatics.jax.org*), many of which being under complex genetic control. The so-called AcB/BcA RCS set was then sought to become a valuable tool to study the genetic basis of many different phenotypes relevant to human health. Ever since the completion of the strain panel and their genetic characterization, this resource has been used by our group and others to study a series of complex phenotypes (Fortin et al 2001a, 2002, Joober et al 2002, Gill & Boyle 2003, 2005a, 2005b, 2005c, Lemay & Haston 2005). Individually, these studies made use of numerous interesting approaches to exploit the strength of RC strains to dissect complex phenotypes. This paper aims to overview such strategies, presenting a series of examples where combinations of mapping and gene expression analyses were used to identify chromosomal regions, genes and biological pathways.

Description of the AcB/BcA recombinant congenic strains

The current set of AcB/BcA consists of 36 RCS, 14 of which having primarily an A/J genetic background, and 22 having primarily a B6 genetic background.

The strains were constructed from the second backcross generation (N3) of the two progenitor mouse strains A/J and C57BL/6J (B6). According to this breeding scheme, each resulting RC strain was predicted to contain a different, randomly distributed set of congenic chromosomal segments corresponding to approximately 12.5% of the donor strain genome, with the complementary 87.5% from the background strain (Demant & Hart 1986). Genotyping the AcB/BcA strain set with 625 microsatellite markers (average coverage of one marker every 2.6 cM) allowed characterization of the genetic composition of individual strains and positioning of their congenic segments (Fortin et al 2001b). Overall, this study revealed that the strain set presented an average donor: background genomic composition of 13%:85% (2% unknown), and an average representation in donor genome content of about 80% on both genetic backgrounds. These results suggested that the number of strains generated together with the distribution of the congenic segments were meeting theoretical expectations, and were suitable for mapping gene-effects.

The AcB/BcA RC strains are generally used to study phenotypes for which A/J and B6 show differential phenotypes. At the beginning of a given study, the parental strains are typically typed or retyped for the trait of interest in order to validate the presence and extent of this differential response, to look at intra-strain variability and to verify the reproducibility of the disease model. Ideally, experimental variability should be reduced to a minimum to facilitate the isolation of individual genetic components. Subsequently, the complete set of RC strains is typed with the disease model to evaluate the genetic complexity of the trait (strain distribution pattern; SDP). In the case of single-gene effects, linkage can readily be established from the SDP of the RC strains (see below). However, for traits controlled by multiple genes, the statistical power within the AcB/BcA RCS set is too limited to allow such direct mapping, and further studies are necessary for unequivocal locus identification. In such case, large segregating secondary crosses (F2) are typically generated from strains showing discordant phenotypes (compared to background-strain) or unique 'extreme' phenotypes. Segregation analysis in these informative strain-derived F2 populations and genetic linkage analyses are subsequently used for locus/loci identification.

Mapping of single gene effects

The ability of the AcB/BcA strains to detect and accurately map a monogenic trait was initially tested in the identification of a locus controlling differential susceptibility to *Legionella pneumophila* (Fortin et al 2001b). It has been established that A/J macrophages are permissive to the invasion and replication by *L. pneumophila* while B6 macrophages can efficiently control their replication. This difference is determined by the expression of a single locus on chromosome 13 designated *Lgn1* (Beckers et al 1995). When infected with *L. pneumophila,* macrophages extracted from the different strains of the AcB/BcA set showed a bimodal distribution of bacterial replication (colony forming unit [CFU] counts), as typically seen for monogenic traits. As shown in Fig. 1, all AcB strains showed A/J-like permissiveness to bacterial replication while most of the BcA strains were resistant with the exception of strains BcA83, 75, 87 and 71, which showed CFU counts in the permissive range. When used as a quantitative phenotypic marker for linkage analysis against the complete genotype set of the RCS, bacterial replication was rapidly linked to a single 3.2 cM genetic interval delineated by marker *D13Mit110* and *D13Mit148*. Since this interval corresponds to the actual position of the *Lgn1* gene (Diez et al 2003, Wright et al 2003), this result clearly illustrates the mapping power of the AcB/BcA RCS set.

Identification and refinement of quantitative trait loci

Despite the important role of quantitative trait locus (QTL) genes in many diseases and knowledge of hundreds of rodent and human QTLs, gene identification by systematic screening of physical intervals underlying QTL regions are difficult due to imprecise delineation and large sizes of the candidate intervals, high number of candidate genes, and fairly low contribution of the QTLs to the trait variance. Recombinant congenic strains can be used to circumvent this problem, as individual susceptibility loci may independently segregate in separate RCS, and the relatively small size of the introgressed congenic segments may facilitate the search and testing of candidate genes. The AcB/BcA RCS was recently used by our group to identify and refine a QTL controlling latency of response to thermal nociceptive stimulus.

The Hargreaves' test of paw withdrawal consists in stimulating the plantar surface of the individual hind paw of a mouse with radiant heat and recording the withdrawal response latency (Hargreaves et al 1988). In a given test session, each hind paw is tested 16 times with at least 5 minutes between successive tests. All animals are subjected to two separate test sessions on different days, and the latencies obtained from the two sessions are averaged. A/J and B6 show differential response to thermal pain stimulation as measured by the Hargreaves' test, with A/J mice showing an average paw withdrawal latency (PWL) of about 10 seconds compared to 5 seconds for B6 (Mogil et al 1999). To identify novel nociception gene(s), the entire set of AcB/BcA RC strains was surveyed using the Hargreaves' test. The goal of this initial step was to identify one or several strains susceptible to have inherited a major nociception controlling locus on an introgressed segment. As seen in Fig. 2 (BcA only), a continuous distribution of the mean PWL was observed within the RCS set, typical of what is expected for multigenic traits. Interestingly, two BcA strains showed a mean PWL very similar to A/J and were

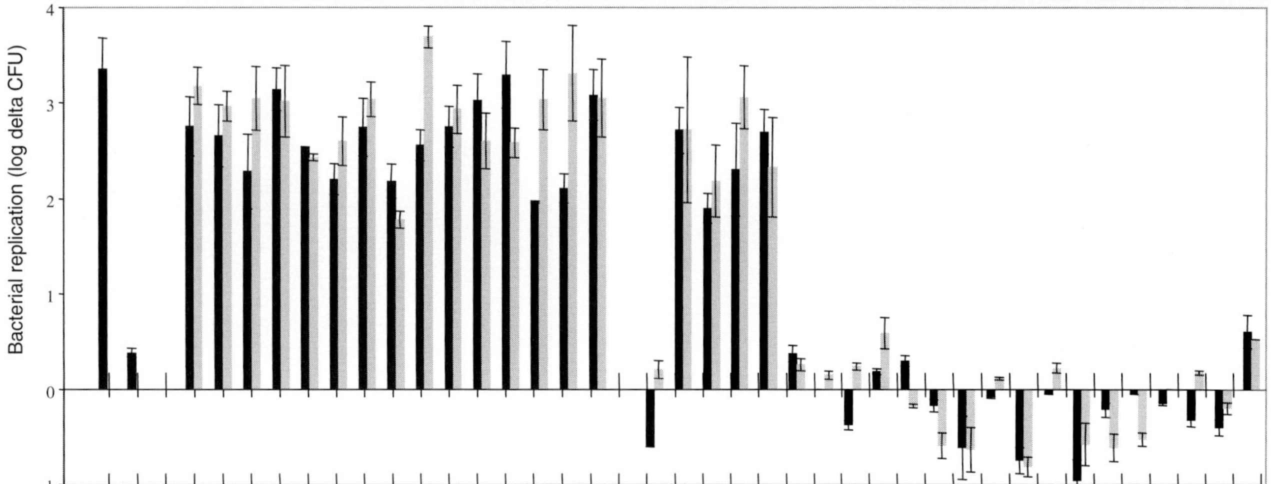

Marker	A	B	AcB54	AcB52	AcB60	AcB65	AcB61	AcB62	AcB59	AcB63	AcB51	AcB56	AcB55	AcB58	AcB53	AcB64	AcB57	BcA79	BcA71	BcA87	BcA83	BcA75	BcA81	BcA72	BcA80	BcA66	BcA74	BcA82	BcA68	BcA76	BcA84	BcA70	BcA85	BcA69	BcA78	BcA86	BcA67	BcA73	BcA77
D13Mit314	A	B	A	A	A	A	A	A	A	A	A	A	A	A	A	A	A	A	A	A	B	B	B	B	B	B	B	B	B	B	B	B	B	B	B	B	B	B	B
D13Mit193	A	B	A	A	A	A	A	A	A	A	A	A	A	A	A	A	A	A	A	A	A	B	B	B	B	B	B	B	B	B	B	B	B	B	B	B	B	B	B
D13Mit106	A	B	A	A	A	A	A	A	A	A	A	A	A	A	A	A	A	A	A	A	A	B	B	B	B	B	B	B	B	B	B	B	B	B	B	B	B	B	B
D13Mit110	A	B	B	A	A	A	A	A	A	A	A	A	A	A	A	A	A	A	A	A	A	A	B	B	B	B	B	B	B	B	B	B	B	B	B	B	B	B	B
D13Die6	A	B	A	A	A	A	A	A	A	A	A	A	A	A	A	A	A	B	A	A	A	A	B	B	B	B	B	B	B	B	B	B	B	B	B	B	B	B	B
Lgn1 →																																							
D13Die3	A	B	A	A	A	A	A	A	A	A	A	A	A	A	A	A	A	B	A	A	A	A	B	B	B	B	B	B	B	B	B	B	B	B	B	B	B	B	B
D13Mit148	A	B	A	B	B	B	A	A	A	A	A	A	A	A	A	A	A	B	A	A	A	B	A	B	B	B	B	B	B	B	B	B	B	B	B	B	B	B	B
D13Mit213	A	B	B	B	B	B	A	A	A	A	A	A	A	A	A	A	A	B	A	A	A	B	A	B	B	B	B	B	B	B	B	B	B	B	B	B	B	B	B
D13Mit292	A	B	A	A	B	B	A	A	A	A	A	A	A	A	A	A	A	B	A	A	A	B	A	B	B	B	B	B	B	B	B	B	B	B	B	B	B	B	B

FIG. 1. Strain distribution pattern of the AcB/BcA RC strains for permissiveness to infection with *Legionella pneumophila*. The permissiveness to *L. pneumophila* replication of the different AcB/BcA RC strains as well as the parental A/J (permissive) and C57BL6/J (non permissive) is presented as bar graphs showing the difference in the number of colony-forming units recovered from infected macrophages 72 h after infection versus that determined immediately after phagocytosis (shown as a logarithm). *L. pneumophila* permissiveness was tested in female (grey bars) and male mice (black bars). Below the bar graphs, the genotypes (A/J, A; C57BL/6J, B) of the different strains at loci surrounding *Lgn1* are shown. Two animals were genotyped for each stain. (Reprinted from Fortin et al 2001b with permission from Elsevier.)

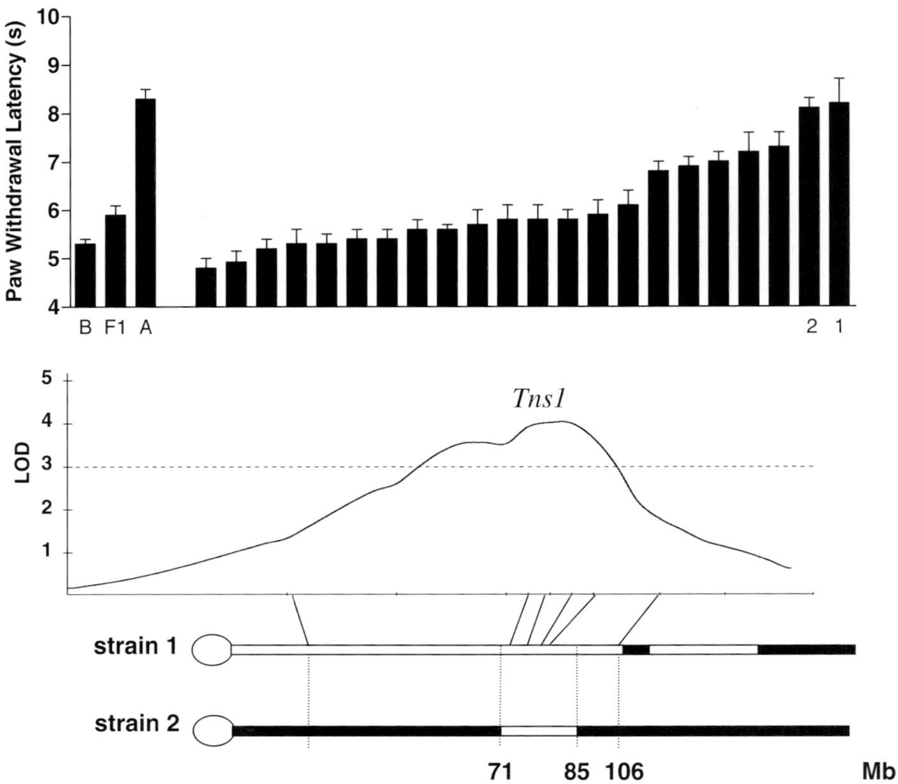

FIG. 2. Identification of the *Tns1* locus. The higher panel presents mean (±S.E.M.) paw withdrawal latencies to noxious thermal stimuli in A/J mice, B6 mice, their F1 hybrids, and in BcA RCS strains. The lower panel presents linkage analysis in F2 hybrids derived from the informative strain 1. Interval mapping was conducted based on the Jackson map with MAPMAKER/EXP version 3.0b and MAPMAKER/QTL version 1.9 using the free model. Schematic representations of the chromosome on which the linkage was identified in both crosses are presented, including the delineation of the selected candidate interval (hatched lines). A/J derived congenic segments and B6 genetic background are represented by white and black parts, respectively.

selected for further genetic studies. Large segregating F2 populations were independently generated from the two informative strains and typed for latency of paw withdrawal using the Hargreaves' test. Subsequently, a limited genome scan procedure using only markers informative for the A/J congenic segments of the informative RC strains (about 15% of total genome, see above) was used for mapping. Interestingly, independent linkage analyses performed on the two F2 populations identified a common significant locus controlling PWL. This locus was named *Tns1* (thermal nociception sensitivity 1).

While the maximum linkage in the cross derived from the first informative strain was detected on a large congenic segment (about 100 Mb), the other strain-derived cross (see Fig. 2) showed the same linkage on an overlapping but smaller congenic segment (14 Mb). Based on the lack of phenotypic complementation between the two informative strains, the smaller A/J-derived introgressed segment was used to physically define the boundaries of our candidate interval. Database mining of the minimal congenic interval identified the presence of 41 genes, a number amenable to systematic candidate screening. In order to identify the gene underlying *Tns1*, expression studies by RT-PCR were initially performed to identify which candidates were expressed in neuronal tissues, selecting twenty of them to be analysed by sequencing. Missense mutations were then identified in the coding regions of six of the remaining candidates. These missense mutations were scrutinized based on the nature of the amino acid change, the position and conservation of the mutated residue, and their presence/absence in unrelated inbred strains of known withdrawal latencies (haplotype analysis). Out of the most interesting mutated candidates, three were prioritized and are currently validated in knockout animals.

Fixing haplotype at known locus/loci while looking for other gene effects

When a major locus is known to control a phenotype of interest, an alternative strategy in using the RCS panel is to select strains with fixed haplotype combinations at this locus to look for novel gene effects. This approach was used by our group to clone the malaria resistance QTL named *Char4* (Fortin et al 2001a). *Plasmodium chabaudi* is the most commonly used parasite to study blood stage malaria in mouse, and A/J and B6 are respectively susceptible and resistant to this infection (Stevenson et al 1982, Yap & Stevenson 1992). QTL mapping by whole genome scanning identified two major loci, termed *Char* (for *Chabaudi resistance*) on central chromosome 8 (*Char*2) and distal chromosome 9 (*Char1*) controlling parasite replication and/or survival to *P. chabaudi* infection (Foote et al 1997, Fortin et al 1997). In the attempt to identify novel malaria resistance genes using the AcB/BcA RCS set, our strategy consisted in looking for strains in the AcB set bearing susceptibility alleles at *Char*1 and *Char*2, but yet showing a malaria resistance phenotype, possibly caused by the transfer of a small B6 chromosomal segment carrying a resistance locus. Following the AcB strain survey, the AcB55 strain was found to be resistant to *P. chabaudi* infection despite susceptibility alleles at *Char1* and *Char2*. Using an informative (AcB55 X A/J) F2 cross, a resistance locus on chromosome 3 (*Char4*) was identified (Fortin et al 2001a).

Further characterization of the AcB55 strain revealed that its resistance to malaria was associated with constitutive reticulocytosis and splenomegaly

(Min-Oo et al 2004). Interestingly, the trait of constitutive reticulocytosis was shown to segregate as a monogenic trait. Genetic mapping using reticulocytosis as phenotypic marker showed that the *Char4* locus was also controlling this trait, and allowed the positioning of the underlying gene more precisely. Subsequently, gene expression using high-density oligonucleotide arrays was used to identify biochemical pathways and cell types underlying the presence of splenomegaly, and allowed candidate searching based on physical location. Sequencing analyses identified a loss-of-function mutation in the pyruvate kinase gene (*Pklr*), which was specific to the informative strain (Min-Oo et al 2003). The *Pklr* mutation was shown to control the phenotype of constitutive reticulocytosis in a bimodal fashion while influencing the outcome of malaria infection in a more complex manner.

Segregation of sub-phenotypes

RC strains can be used to segregate different sub-phenotypes present in the parental strains, as illustrated in the following example. When measured by Piximus imaging system, the total amount of trabecular bone, as translated by bone mineral density (BMD), is significantly lower in the femur and vertebra of A/J mice compared to B6. This is corroborated by micro computed tomography (CT) and histomorphometric analyses showing lower bone volume on tissue volume (BV/TV) in the A/J mice. The architecture of trabecular bone is also compromised in A/J mice, as shown by an increase in the structure model index (SMI) that compares plate-like to rod-like structures. However, the thickness of the cortical bone in the diaphyseal region is greater in A/J mice compared to B6. In a recent study, the AcB/BcA RC strains were used by our group as a discovery platform to identify genes controlling bone formation. Among the BcAs, one strain was identified as showing pheno-deviance in trabecular bone formation consistent with the A/J phenotype. Indeed, this informative strain showed decreased BMD and BV/TV, increased SMI and decreased bone apposition and turnover despite a primarily B6 genetic background. Figure 3 shows representative images of femur and vertebra from A/J, B6 and from the informative strain. Interestingly in the pheno-deviant stain, the trabecular phenotype is similar to A/J while the cortical phenotype remains similar to the B6 parent.

Segregation analysis of eQTLs in F2 mice to define pathways

Several epidemiological studies have shown that the concentration of plasma high-density lipoprotein (HDL) correlates inversely with the risk of cardiovascular disease (CVD) and that HDL level is a powerful predictor of CVD development (Gordon & Rifkind 1989). In the attempt of identifying genes involved in lipid

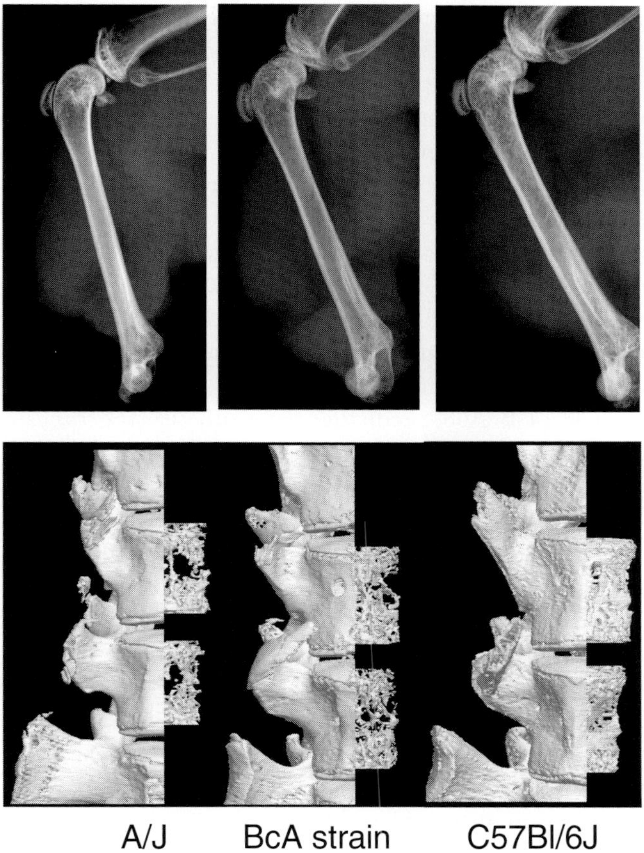

FIG. 3. Cortical and trabecular bone compartments of A/J, C57BL/6J and of one informative BcA strain. Faxitron X-ray images (upper panel) and 3D reconstruction from high resolution micro computed tomography (MicroCT) scans (lower panel) were used to visualize the cortical and trabecular bone compartments of the femur and the 4th and 5th lumbar vertebrae of each strain, respectively.

metabolism in mice, the AcB/BcA RCS along with parental controls were fed a high-fat 7.5% cacao butter diet for a period of 4 weeks. Lipid fractions were measured before and after the diet period, and strain distribution patterns were analysed. Amongst all other strains including the parentals, BcA68 presented a unique phenotype of extremely low levels of HDL-c, total cholesterol and triglycerides. This distinct phenotype was present in males and females, and was not greatly affected by the high-fat diet. Together these data suggested that the phenotype seen in BcA68 was due to a *de novo* mutation, not necessarily present on a congenic segment. Segregation analyses using F1 and F2 mice derived from

BcA68 confirmed that the phenotype of extremely low HDL-c was segregating as a single recessive autosomal trait. To map the causal gene, we chose to cross BcA68 with unrelated inbred strains, AKR and DBA/2, to generate hybrids in which the complete genome was informative. A low-coverage genome survey (40 SSLP markers) was undertaken on a group of 21 low HDL-c (BcA68xAKR)F2 animals along with 21 normal HDL-c control littermates. Using this mapping strategy, the phenotype of low HDL-c was rapidly linked to a locus on chromosome 9 encoding a cluster of genes previously shown to be involved in lipid metabolism, including *ApoA1*, *ApoA4*, *ApoC3*, *ApoA5* and *Pafah1b2*. Since ApoA1 deficiency was known to cause extremely low levels of HDL-c in humans and a genetically engineered mouse model, this gene was prioritized for sequence analysis. Two consecutive point mutations were found in exon 3 of the *ApoA1* gene (66–67 GC > TT) of BcA68, predicting the synthesis of a truncated 23 amino acids ApoA1 protein in the BcA68 RC strain.

In order to identify novel genes involved in the same biological pathway as *ApoA1*, we performed high-density oligonucleotide arrays and quantitative-PCR on RNA extracted from livers of parental A/J and B6, BcA68 as well as of (BcA68XB6)F2 mice bearing or not the *ApoA1* loss of function mutation, and submitted or not to a high fat diet. Interestingly, the mutation found in *ApoA1* correlated with a four-fold lower expression of the gene in normal (Fig. 4) and high fat diet-fed animals (not shown). In addition, we observed in the four groups of F2 that the level of expression of *ApoA1* correlated perfectly with the level of plasma HDL-c. Using cluster analysis, our goal is now to identify genes showing a similar pattern of expression as *ApoA1*, which will be suggestive of a potential role in lipid metabolism in general, or in HDL formation in particular.

In summary, the AcB/BcA set of recombinant congenic strains has proven to be a powerful and versatile genetic tool to dissect several complex phenotypes including nociception, malaria susceptibility, bone formation and lipid metabolism, and to identify their genetic components. Our results show that RCS are useful to rapidly identify single gene effects, map and refine QTL intervals, dissect sub-phenotypes, and may on occasion provide a novel source of genetic diversity in the form of *de novo* mutation fixed by inbreeding during the production of the strains. As the phenotyping data on the RCS set continues to grow, it should become possible to identify strain effects, including specific genotypes and eQTLs that may commonly regulate responses to different stimuli. This may allow the discovery of unexpected connections between complex diseases, possibly revealing common mechanisms of pathogenesis.

Acknowledgements

The AcB/BcA set of RCS is supported by a grant from Genome Quebec/Genome Canada.

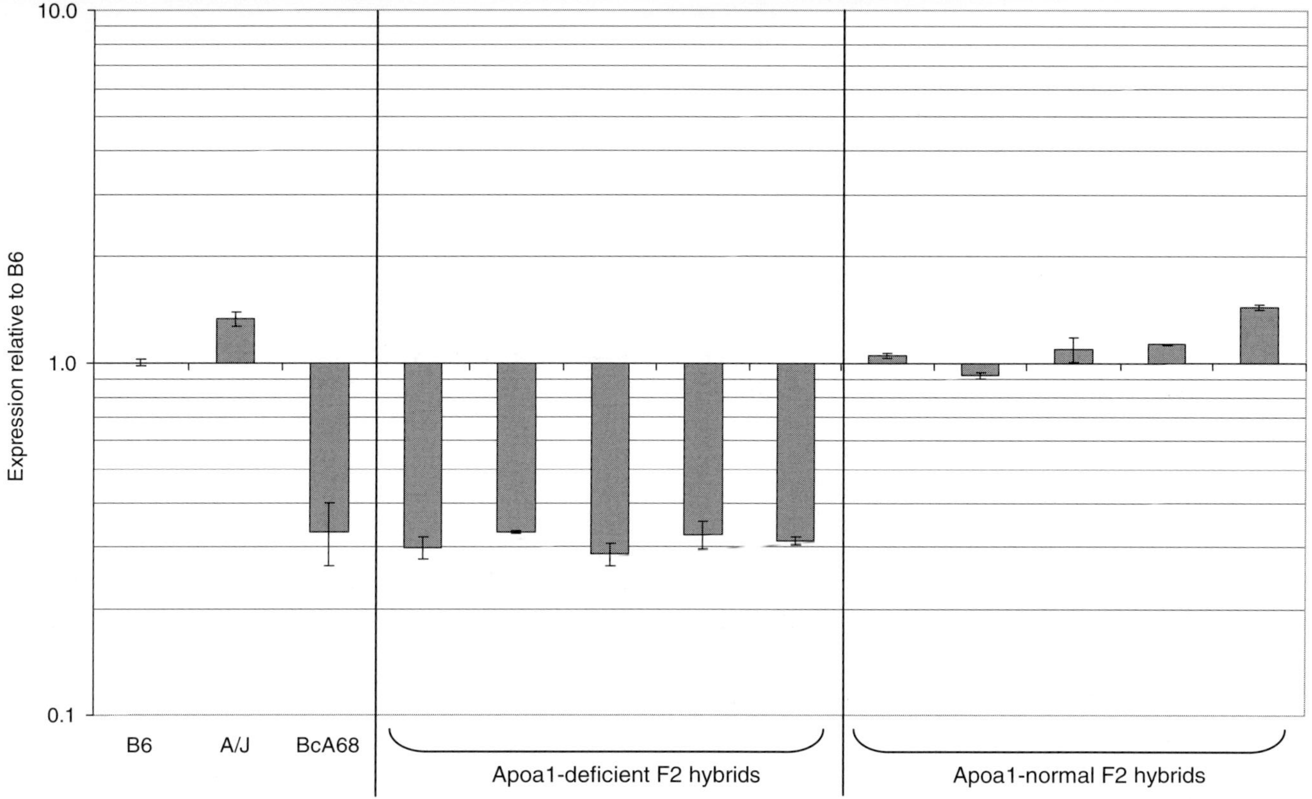

FIG. 4. Effect of the ApoA1$^{66-67TT}$ truncation mutation on liver expression of the *ApoA1* transcripts. Liver cDNAs were prepared from male A/J, C57BL/6J and BcA68 inbred mouse strains as well as from five ApoA1-deficient (*ApoA1*$^{66-67TT}$) and five ApoA1-normal (*ApoA1*$^{66-67GC}$) (BcA68 x C57BL/6J) F2 mice. Quantitative PCR was performed using Taqman assays. Four replicate samples were analysed from each of two independent RNA extractions. *Apoa1* gene expression levels were normalized to the median of three housekeeping genes tested (*Actin*, *Pgk1* and *Hprt1*). Normalized transcript levels in all tissues were arbitrarily expressed relative to those detected in the C57BL/6J strain (±SEM).

References

Balling R 2001 ENU mutagenesis: analyzing gene function in mice. Annu Rev Genomics Hum Genet 2:463–492

Beckers MC, Yoshida S, Morgan K, Skamene E, Gros P 1995 Natural resistance to infection with Legionella pneumophila: chromosomal localization of the Lgn1 susceptibility gene. Mamm Genome 6:540–545

Clark AT, Goldowitz D, Takahashi JS et al 2004 Implementing large-scale ENU mutagenesis screens in North America. Genetica 122:51–64

Darvasi A, Soller M 1995 Advanced intercross lines, an experimental population for fine genetic mapping. Genetics 141:1199–1207

Demant P, Hart AA 1986 Recombinant congenic strains—a new tool for analyzing genetic traits determined by more than one gene. Immunogenetics 24:416–422

Diez E, Lee SH, Gauthier S et al 2003 Bircle is the gene within the Lgn1 locus associated with resistance to Legionella pneumophila. Nat Genet 33:55–60

Foote SJ, Burt RA, Baldwin TM et al 1997 Mouse loci for malaria-induced mortality and the control of parasitaemia. Nat Genet 17:380–381

Fortin A, Belouchi A, Tam MF et al 1997 Genetic control of blood parasitaemia in mouse malaria maps to chromosome 8. Nat Genet 17:382–383

Fortin A, Cardon LR, Tam M, Skamene E, Stevenson MM, Gros P 2001a Identification of a new malaria susceptibility locus (Char4) in recombinant congenic strains of mice. Proc Natl Acad Sci USA 98:10793–10798

Fortin A, Diez E, Rochefort D et al 2001b Recombinant congenic strains derived from A/J and C57BL/6J: a tool for genetic dissection of complex traits. Genomics 74:21–35

Fortin A, Stevenson MM, Gros P 2002 Complex genetic control of susceptibility to malaria in mice. Genes Immun 3:177–186

Gill KJ, Boyle AE 2003 Confirmation of quantitative trait loci for cocaine-induced activation in the AcB/BcA series of recombinant congenic strains. Pharmacogenetics 13:329–338

Gill KJ, Boyle AE 2005a Genetic analysis of alcohol intake in recombinant inbred and congenic strains derived from A/J and C57BL/6J progenitors. Mamm Genome 16:319–331

Gill KJ, Boyle AE 2005b Genetic basis for the psychostimulant effects of nicotine: a quantitative trait locus analysis in AcB/BcA recombinant congenic mice. Genes Brain Behav 4:401–411

Gill KJ, Boyle AE 2005c Quantitative trait loci for novelty/stress-induced locomotor activation in recombinant inbred (RI) and recombinant congenic (RC) strains of mice. Behav Brain Res 161:113–124

Gordon DJ, Rifkind BM 1989 High-density lipoprotein—the clinical implications of recent studies. N Engl J Med 321:1311–1316

Hargreaves K, Dubner R, Brown F, Flores C, Joris J 1988 A new and sensitive method for measuring thermal nociception in cutaneous hyperalgesia. Pain 32:77–88

Joober R, Zarate JM, Rouleau GA, Skamene E, Boksa P 2002 Provisional mapping of quantitative trait loci modulating the acoustic startle response and prepulse inhibition of acoustic startle. Neuropsychopharmacology 27:765–781

Lemay AM, Haston CK 2005 Bleomycin-induced pulmonary fibrosis susceptibility genes in AcB/BcA recombinant congenic mice. Physiol Genomics 23:54–61

Min-Oo G, Fortin A, Tam MF, Nantel A, Stevenson MM, Gros P 2003 Pyruvate kinase deficiency in mice protects against malaria. Nat Genet 35:357–362

Min-Oo G, Fortin A, Tam MF, Gros P, Stevenson MM 2004 Phenotypic expression of pyruvate kinase deficiency and protection against malaria in a mouse model. Genes Immun 5:168–175

Mogil JS, Wilson SG, Bon K et al 1999 Heritability of nociception I: responses of 11 inbred mouse strains on 12 measures of nociception. Pain 80:67–82

Singer JB, Hill AE, Burrage LC et al 2004 Genetic dissection of complex traits with chromosome substitution strains of mice. Science 304:445–448
Stevenson MM, Lyanga JJ, Skamene E 1982 Murine malaria: genetic control of resistance to Plasmodium chabaudi. Infect Immun 38:80–88
Wright EK, Goodart SA, Growney JD et al 2003 Naip5 affects host susceptibility to the intracellular pathogen Legionella pneumophila. Curr Biol 13:27–36
Yap GS, Stevenson MM 1992 Plasmodium chabaudi AS: erythropoietic responses during infection in resistant and susceptible mice. Exp Parasitol 75:340–352

DISCUSSION

Foote: The pyruvate kinase mutation was also a *de novo* mutation, so the two you have identified were essentially *de novo* mutations that you would have been able to identify even in the absence of having these recombinant strains. Is that correct?

Fortin: It is correct that for both the mutations detected in the *ApoA1* and the *pyruvate kinase* genes, the recombinant congenic strains set provided a source of new mutations. This does not show that the RCS tool is not efficient to identify parentally-derived QTL genes, but reflects the fact that for strong genetic effect, single gene mutations may be the first to be identified in a RCS screen for a phenotype of interest. If you look at the nociception project, our results suggest that the causative mutation has been inherited from the A/J parental strain, meaning that in this case, the strain set has allowed the isolation of a parentally derived QTL.

Morahan: What is your theoretical resolution across the panel of congenic strains?

Fortin: Calculating the theoretical mapping resolution of the RCS set is difficult for complex traits as the congenic segments are of variable sizes and the strength of 'mappable' gene effect may vary dramatically. As for simple traits, our past experience suggests a resolution of about 5 to 15 Mb.

Goodnow: Why do you need to do the breeding? I didn't quite follow that.

Fortin: Mapping a simple trait using the RC strains is efficient as there is usually enough power within the 36 RC strains to identify the causing locus. However, this is not the case for multigenic traits. Consequently, secondary F2 crosses involving the strain(s) of interest need to be generated to asses the effect of the individual congenic segments.

Vinuesa: How generalizable is this finding of monogenic traits within complex traits? Is this the key to being able to map most of these complex traits?

Fortin: I have no answer for this question. However, I do believe that in the context of a complex trait, maximal information on the phenotype is a key point, as it can strengthen the power of the genetic analysis and shed light on the selection of relevant candidates during the gene identification process.

Wakeland: Of all the traits for which you had a nice, robust difference between the parentals, which you then tried to map in your RC strain combination, how many panned out in a way that you could then proceed?

Fortin: As of now we have typed the RC strains for five different traits in house and have pursued all of them by mapping in secondary crosses. As for the gene identification step, only congenic segments associated with significant linkages are further studied. So far, we have preferred an approach in which all the genes delineated by the boundaries of a congenic segment are systematically screened regardless of their positions relative to the peak linkage. Priorization of the candidates is nevertheless performed based on potential relevance to the phenotype of interest. However, this strategy could be revised depending on the size of the congenic segment and the number of genes in the region. In the context of the nociception project, we had the advantage of having a second informative strain to help refine the QTL region. In addition, the final candidate interval was relatively gene-poor.

Wakeland: If you do an F2 of the parental screen, you are in essence in a position where epistatic interactions are likely to be avoided, because you have a much bigger dose of the parental genome. I am on the study section of the NIH, and I spend a lot of time reading proposals for using RI strains inappropriately, with people hoping to find QTLs simply by typing the RI strains. This never works. It is nice to see a well thought through strategy that works nicely.

Fortin: It is true that mapping QTLs directly in the RCS only works for simple strong monogenic effects, as discussed before. Our mapping strategy is based on previous experience with gene identification in general, and on what we have seen so far in this particular set of RCS.

Goodnow: Grant Morahan, I know you are embarking on a big adventure along these lines. How does this fit into your thinking?

Morahan: We are setting up what we call 'The Collaborative Cross', which is a series of 1000 RI strains, each descended from eight 'great-grandparental' founder strains (Churchill et al 2005). These founder strains were selected to be the most diverse set among the inbred strains available worldwide. 1000 'cousin' strains should give sufficient resolution to map single-gene traits down to the individual gene level, and it should also generate sufficient power to be able to map or identify three or four interacting genes. How it differs from classical genetic technologies is that most of the breeding and characterization is done up-front. In order to map their gene, researchers will just need to test the panel for their trait is of interest, and compare the pattern of that trait with the pattern of genetic markers. This has been pretty well worked out. We currently have 400 funnels set up, with 400 more strains being bred at Oak Ridge National Laboratory (Robert Williams and colleagues) and another 200 in Kenya (Richard Mott and Fouad Iraqi).

Goodnow: Do you think that just by the number of strains you will overcome this need to have to do a second cross?

Morahan: It depends on the genetic architecture. For single gene traits we won't need the second cross. For more complex traits it depends on the effect size of each gene and the number of genes in the mix.

Gros: In that cross, are you taking any precaution to minimize the propagation and fixing of *de novo* mutations? These are the ones that will come and surprise you. They jump up as being the most exciting phenotype, yet they don't rely on the set-up of the cross, or they don't exploit the conceptual framework that drove the construction of the strain.

Morahan: We have thought about this. It is a potential bugbear of this project. If you get spontaneous mutants you are back into a normal Mendelian experiment to map and identify them. There's no getting around it, other than making sure your mice breathe nice pure West Australian air! We are also building in some failsafe provisions by cryopreserving embryos at key generations in the breeding program.

Reference

Churchill GA, Airey DC, Allayee H et al 2004 The Collaborative Cross, a community resource for the genetic analysis of complex traits. Nat Genet 36:1133–1137

Genetic control of host–pathogen interactions in mice

Gundula Min-Oo*, Mary M. Stevenson†, Anny Fortin‡* and Philippe Gros†

**Department of Biochemistry and †Center for the Study of Host Resistance, McGill University, Montreal, ‡Emerillon Therapeutics, Montreal, Canada*

Abstract. The onset, progression and outcome of infections are determined by performance of host defence mechanisms and expression of pathogen virulence determinants. Genetic analysis in mouse can identify host genes that play critical roles at the interface of host–pathogen interactions. Genetic effects detected as variations in susceptibility in inbred, recombinant and mutant strains of mice can be mapped as simple traits or quantitative trait loci followed by identification by positional cloning. We have used mouse models of infection with bacterial (*Mycobacterium, Legionella*) and parasitic pathogens (*Plasmodium*) to discover genes and proteins that are important for macrophage function against such infectious agents. These studies have identified Nramp1-mediated exclusion of divalent metals from the phagosomal space as a key regulator of intracellular replication of *Mycobacteria*. Also, intracellular sensing of *Legionella* by functional Birc1e/Naip5 protein is essential to prevent replication of this bacterium in macrophages. Finally, we have identified two new loci that affect blood-stage replication of *Plasmodium chabaudi* AS in mice, and have cloned the corresponding genes.

2007 Decoding the genomic control of immune reactions. Wiley, Chichester (Novartis Foundation Symposium 281) p 156–168

The apparent heritability of susceptibility to infectious diseases has long been recognized in humans, with diseases such as tuberculosis and leprosy being thought to 'run' in certain families (Cooke & Hill 2001). One of the clearest examples of such a genetic effect is the case of haemoglobinopathies, including sickle cell anaemia, β-thalassemias and glucose-6-phosphate dehydrogenase deficiency, where heterozygosity for otherwise-deleterious alleles has a protective effect against blood-stage replication of the malarial parasite *Plasmodium falciparum* (Min-Oo & Gros 2005). Other clear examples include the presence of independent mutations in the interferon (IFN)γ/interleukin (IL)12 host response pathway in patients developing disseminated mycobacterial infection following BCG vaccination (Casanova & Abel 2004). However, such instances of single gene effects are relatively rare, and the genetic component of susceptibility to infections is often complex, multigenic and further modified by environmental effects including health status of the host and variable pathogen-associated virulence determinants (Hill 1998). Nevertheless, identifying genes affecting initial susceptibility, progression, host response and ultimate outcome of infections may reveal host proteins playing a key role in pathogenesis including natural or acquired immune defences against a specific pathogen.

Taking advantage of the genetic diversity found in commercially available inbred mouse stocks and within a set of AcB/BcA recombinant congenic mouse strains (Fortin et al 2001a), we have used experimental infections to identify loci that affect susceptibility to infection with different intracellular pathogens. These loci have been detected as monogenic traits or as more complex quantitative trait loci (QTLs). A whole-genome scanning approach in informative crosses was used to locate major contributing loci, which in some instances have been identified by positional cloning and validated by the creation of loss-of-function mutations, by gene targeting or gain-of-function mutations in transgenic animals. In addition, complex genomic regions have been studied by functional complementation using recombinant clones (BAC) from the region transferred into transgenic lines. We will herein review representative examples of positional cloning from our lab, with a focus on recent studies on genetic control of blood-stage replication of the malarial parasite.

Simple traits

Single gene effects accounting for a large fraction of the phenotypic variance in an informative cross are most amenable to positional cloning (Lam-Yuk-Tseung & Gros 2003). Although whole genome sequence information has greatly facilitated this endeavour, gene identification and validation remains a challenging task. Examples of novel insight into host–pathogen interactions gained by the successful cloning of single gene effects will be reviewed.

Susceptibility to infection with mycobacterial species in mice (as measured by early *in vivo* microbial replication in spleen) is controlled by a locus on chromosome 1 termed *Bcg*. *Bcg* was found to be tightly linked or allelic with the *Ity* and *Lsh* loci previously shown to regulate *in vivo* replication of *Salmonella typhimurium* and *Leishmania donovani* (Skamene et al 1982, 1998). Studies in explanted cell populations showed that *Bcg/Ity/Lsh* affects the ability of macrophages to restrict intracellular replication of these unrelated pathogens. The cloning and subsequent functional characterization of *Nramp1* as the gene responsible for *Bcg/Ity/Lsh* represents one of the earliest demonstrations of the power of genetics to identify key host response mechanisms to infection (Skamene et al 1998). In the absence of a mouse genome sequence, positional cloning involved (a) the construction of high resolution genetic (0.2 cM) and physical maps (475 kb) of the region, (b) the localization of transcribed genes based on clustering of CpG islands on the physical map, (c) the identification of one of them (*Nramp1*, renamed *Slc11a1*) expressed exclusively in phagocytes, (d) the association of a unique haplotype/mutation (G169D) with susceptibility to infections in inbred strains, and finally (e) validation in mutant mice bearing loss (knockout) or gain of function (transgenic G169 allele) mutations at *Nramp1* (Forbes & Gros 2001). Nramp1 is a membrane protein composed of 12 transmembrane domains that is expressed specifically in Lamp1$^+$ lysosomes of macrophages and in the gelatinase$^+$ tertiary granules of neutrophils. Upon phagocytosis of pathogens by these phagocytes Nramp1 is recruited to the membrane of the phagosome. Biochemical studies with metal-sensitive fluorescent dyes coupled to zymosan particles have shown that Nramp1 functions at the phagosomal membrane as a pH-dependent efflux pump for divalent metals including Fe^{2+} and Mn^{2+} (Forbes & Gros 2001). Metal chelation by Nramp1 has pleiotropic effects on the metabolic activity of intra-phagosomal pathogens that includes antagonizing intracellular survival strategies developed by *Mycobacterium* (reduced acidification, and inhibition of fusion to lysosomes) (Frehel et al 2002) and *Salmonella* (reduction of recruitment of manose-6-phospate receptor and fusion with early endosomes) (Cuellar-Mata et al 2002). *Salmonella* responds to Nramp1-induced metal deprivation by activating virulence factors encoded by the specific pathogenicity island I including metal transport system (Zaharik et al 2002). Finally, an association between polymorphic variants at the human *NRAMP1* gene and susceptibility to tuberculosis, leprosy, HIV, and other infectious and autoimmune diseases has been noted (Poon & Schurr 2004). Together, these studies indicate that competition for divalent metals in the phagosomal space plays a critical role in host–pathogen interactions.

Positional cloning is sometimes impeded by the complex nature of the genomic region involved, requiring use of additional genomic tools. Susceptibility to infection with mouse cytomegalovirus (MCMV) is controlled by a single locus, *Cmv1*, on chromosome 6. The 1.6 Mb *Cmv1* physical interval contains 18 transcription units, including the 14 members of the *Ly49* family (Lee et al 2003a, Webb et al 2002). Ly49 members are cell surface receptors expressed on overlapping subsets of natural killer (NK) cells, that bind class I MHC molecules and transmit activating and inhibitory signals to direct cell-mediated cytotoxicity against virally-infected cells. While studying the unique MCMV susceptibility of the BXD8 mouse strain (despite presence of resistant alleles for *Cmv1*^{R}), we discovered a 23 kb deletion in the *Cmv1* region that encompasses the *Ly49h* gene, strongly suggesting that *Cmv1* and *Ly49h* are allelic (Lee et al 2001). However, independent high resolution mapping of *Cmv1* to non-overlapping 23 kb and 220 kb subregions of the 1.6 Mb interval suggested possible genetic heterogeneity in the effect of *Ly49* genes in resistance to MCMV. This was evaluated *in vivo* by functional complementation using BAC transgenesis (Lee et al 2003b). BAC clones from C57BL/6J (*Cmv1*^{R}), and overlapping either the 23 kb proximal (128D23), or the 200 kb distal (13J11) candidate region, were introduced into FVB/N zygotes and their ability to correct MCMV susceptibility was tested after backcrossing the transgenic BAC to BALB/c (*Cmv1*^{S}). *Ly49h*-containing BAC 128D23, but not 13J11, was shown to correct MCMV susceptibility, and the extent of complementation afforded by this clone was proportional to the level of cell surface expression of the transgenic Ly49H protein on recipient NK cells (Lee et al 2003b). These BAC transgenesis experiments confirmed that *Cmv1* and *Ly49h* are indeed allelic.

Inbred mouse strains are resistant to infection with the intracellular pathogen *Legionella pneumophila*, with the notable exception of A/J macrophages. The unique susceptibility of A/J is controlled by a single locus on chromosome 13: *Lgn1* (Fortier et al 2005). The *Lgn1* interval was delineated to a 0.31 cM segment that displays a direct repeat structure, including multiple copies of the *Birc1* gene (baculoviral inhibitor of apoptosis protein repeat-containing 1; formerly *Naip*) (Growney & Dietrich 2000). Amongst resistant strains (*Lgn1*^{r}), this region contains a variable number of additional rearranged copies of *Birc1*, precluding standard positional cloning approach to pinpoint which *Birc1* copy underlies *Lgn1*. *In vivo* complementation with BAC transgenics was used to answer this question. Transgenic FVB mice were created by injecting four BAC clones containing discrete portions of the *Birc1* cluster from resistant B6 mice followed by backcrossing of the transgene onto the susceptible A/J background (Diez et al 2003). RNA was isolated from macrophages from these mice and analysed for transgenic *Naip* copies mRNA expression by RT-PCR, using strain-specific SNPs that could distinguish between A/J, B6 and FVB-derived *Birc1* transcripts. Testing for complementation of the A/J-derived susceptibility phenotype by *ex vivo* infection of peritoneal macrophages indicated that the two BAC clones that rescued the phenotype both had the full length *Birc1e* transcript in common (Diez et al 2003). These results showed that *Birc1e* is likely allelic with *Lgn1*. *Birc1* protein(s) is expressed in macrophages, and it is up-regulated upon phagocytosis of infectious agents or inert particles. Birc1e/Naip5 were recently shown to belong to the NLR protein family, a group of cytoplasmic proteins involved in intracellular recognition of conserved structures called pathogen-associated molecular patterns. NLR proteins show a conserved nucleotide-binding domain mediating protein dimerization, a conserved leucine-rich repeat responsible for ligand recognition, and an intracellular signalling module (Inohara et al 2005). Recent studies have shown that Birc1e/Naip5 can mediate caspase 1 induced cell death and IL1 production in response to *Legionella* products; Furthermore, bacterial products transported by the type IV secretion system Dot-Icm are required to reveal the Birc1e/Naip5 effect (Zamboni et al 2006).

Complex traits

The single gene effects discussed above provide clear examples of how genetic analysis in mouse models can identify novel host proteins and pathways playing important roles in host–pathogen interactions. In many instances, the genetic component of susceptibility is complex and multigenic, with individual gene effects mapped as QTLs contributing only a small proportion of the phenotypic variance. In addition, the size of chromosomal segments harbouring these QTLs is often so large that isolating potential candidates from the region is difficult and requires additional genomic approaches. Our laboratory has used recombinant congenic mouse strains (Fortin et al 2001a) to isolate single gene effects contributing to complex traits. In addition, transcript profiling with microarrays has proven valuable for identifying positional candidates in large chromosomal regions underlying individual QTLs. This approach has been applied to the study of the complex genetic control of susceptibility to malaria.

Chabaudi resistance (Char) loci

Among the murine malaria parasites (*P. chabaudi*, *P. yoelii*, *P. berghei*, *P. vinckei*), *P. chabaudi* AS provides a unique experimental tool with many similarities to the human parasite *P. falciparum*. The symptoms and pathology, including anaemia, associated with blood-stage *P. chabaudi* AS infection in mice correspond to a reasonable degree to those associated with *P. falciparum* malaria in humans (Li et al 2001, Stevenson & Riley 2004), making this model useful to study pathophysiology and define parameters of host response in human. Our laboratory has studied the genetic factors that regulate blood-stage replication of the malarial parasite. In inbred strains, A/J, BALB/c, C3H/HeJ and 129/ICR are susceptible while C57BL6/J and DBA/2 are resistant to *P. chabaudi* AS infection. In general F1 hybrids are more resistant than either parents, indicating that resistance behaves as dominant, and suggesting complementation of additional parental susceptibility loci in F1 mice (Fortin et al 2002). When infected with *P. chabaudi*-parasitized erythrocytes, A/J mice die of progressive disease while B6 mice overcome the infection and clear the parasite load within two weeks. The genetic advantage of B6 is expressed as (1) reduced parasitaemia at the peak of infection, (2) increased inflammatory response, (3) increased TNFα production, and (4) survival to infection. Studies in 16 AXB/BXA recombinant inbred strains derived from B6 and A/J parents have shown that genetic susceptibility is complex and multigenic (Fortin et al 2002). QTL mapping by whole genome scan in informative [A x B6] backcross and F2 mice by our group (Fortin et al 1997) and in SJL x B6 and C3H x B6 F2 mice by Foote et al (1997) mapped 2 major *Chabaudi resistance* loci on chromosomes 9 (*Char1*) and 8 (*Char2*) that control peak parasitaemia during *P. chabaudi* infection. A third H-2-linked locus on chromosome 17, (*Char3*) was found to regulate parasite clearance immediately after the peak of infection (Burt et al 1999). However, these three loci individually account for a modest fraction of the phenotyptic variance, and the large (20–30 cM) size of their genetic interval renders difficult their identification by positional cloning. Recently, four additional suggestive linkages (LOD < 2.0) have been detected in F11 advanced inter-cross lines (from A and B6 parents) that have a modest effect on blood stage replication of *P. chabaudi*, and that map to chromosomes 5 (*Char5*, *Char6*), 17 (*Char7* related to *Char3*) and 11 (*Char8* syntenic with human 5q31–q33) (Hernandez-Valladares et al 2004). So far, none of the genes involved have been cloned.

Char4

Our laboratory has used AcB/BcA recombinant congenic strains (RCS) bred from malaria-susceptible A, and resistant B6 parents (Fortin et al 2001a; see chapter by Fortin et al 2006, this volume) to detect novel gene effects affecting blood stage replication of *P. chabaudi AS* (Fortin et al 2002). By virtue of the breeding scheme used in their derivation, individual RCS contain a small amount (12.5%) of DNA from one parent fixed as a set of discrete congenic segments, on the genetic background (87.5%) of the other parent. Therefore, individual resistance/susceptibility loci may independently segregate in individual RCS. The relatively small size of the congenic segments fixed in individual RCS facilitates the search and testing of candidate genes. Although a good correlation was detected between resistance/susceptibility to *P. chabaudi* infection and haplotypes on chromosomes 8 (*Char2*) and 9 (*Char1*) in AcB/BcA strains, strains AcB55 and AcB61 were discordant, being very resistant to malaria despite susceptibility alleles at *Char1* and *Char2*. In both strains, resistance was associated with reduced peak parasitaemia, and early clearance of infection. Linkage analysis in a (AcB55 X A)F2 mice population located a new B6-associated resistance locus on chromosome 3, that we designated *Char4* (Fortin et al 2001b). Resistance in AcB55 mice was associated with splenomegaly, elevated levels of Ter-119$^+$ erythrocyte precursors in this organ, as well as extramedullary erythropoiesis in liver. Transcriptional profiling with cDNA arrays and using RNA from spleen and liver of AcB55 vs. A mice identified increased expression of several erythroid-specific genes in AcB55. Such a profile was correlated with a constitutive reticulocytosis noted in peripheral blood of AcB55. Interestingly, similar transcriptional profiles in spleen and reticulocytosis in peripheral blood were noted in AcB61. Re-examination of the [AcB55 x A]F2 and [AcB61 x A]F2 animals using circulating reticulocyte counts indicated unigenic mode of inheritance for reticulocytosis. Complementation studies in compound [AcB61 x AcB55] F1, together with linkage studies in informative F2 indicated that the locus mapped to the proximal portion of chromosome 3, in a region devoid of B6 parental genomic DNA, suggesting that the trait appeared as a result of a new mutation emerging on A/J background, and becoming fixed during the breeding of the AcB55 and AcB61 strains. Combined analysis of (a) the transcript map of the candidate chromosomal region for presence of genes expressed in erythroid cells, and of (b) the list of genes differentially expressed in AcB55 vs. A tissues, identified the liver/erythrocyte-specific isoform of pyruvate kinase (PK) as a likely candidate (*Pklr*). This gene was sequenced and a single mutation (I90N) was identified in the PK gene from both AcB55 and AcB61 (Min-Oo et al 2003, Min-Oo & Gros 2005). A direct correlation was detected in 400 informative [AcB55 x A] and [AcB61 x A]F2 mice between $PKLR^{I90N}$, PK enzyme activity, reticulocyte counts, peak parasitaemia and survival following *P. chabaudi* infection. These findings demonstrate that loss-of-function at the erythrocyte form of PK has a protective effect against blood stage replication of the malarial parasite, including associated mortality (Min-Oo et al 2003). We have recently obtained a second *Pklr* mutant allele (G338D) from the CBA-$Pk1^{slc}$ mouse. CBA $Pklr^{G338D}$ mice were found to be extremely resistant to infection with *P. chabaudi* AS with very low peak parasitaemia and no mortality compared to CBA/N and CBA/J controls, confirming the protective effect of PK deficiency on susceptibility to malaria in mice.

Char9

During the initial mapping of *Char4*, a second weaker linkage was detected on chromosome 10 in the [AcB55 x A]F2 cross (Fortin et al 2001b). Linkage was to a B6-derived chromosome 10 segment of ~15 Mb in size, with maximal effect detected with marker *D10Mit189* (LOD = 3.0). The B6 alleles at this locus (given the temporary designation *Char9*) are protective against *P. chabaudi* AS infection (reduced peak parasitaemia), and inherited in a co-dominant fashion (Fortin et al 2001b). Because the marker initially used to map the *Char4* effect (*D3Mit109*) is 20 cM distal to the actual causative *Pklr* mutation, the *Char9* linkage data was re-analysed after fixing wild-type and mutant *Pklr* ($Pklr^{I90N}$) alleles segregating in this F2 cross. This analysis identified a robust effect for *Char9* (LOD = 4.7) in the central portion of the B6-derived chromosome 10 congenic segment fixed in AcB55. *Char9* alleles modulate peak parasitaemia on all 3 *Pklr* genotypes. The minimal genetic interval of the B6-derived chromosome 10 segment harbouring *Char9* in AcB55 contains 82 genes that have been systematically characterized with respect to (a) tissue-specific mRNA expression (transcript profiling with microarrays), (b) strain-specific variations in mRNA expression (quantitative RT-PCR), and (c) sequence polymorphisms associated with differential susceptibility. These studies have identified a strong candidate for *Char9*. This gene is the only transcript in the *Char9* region that is expressed in the spleen and liver of resistant C57BL/6J and AcB55 strain but is absent from susceptible A/J and AcB61. Sequence analyses identify two transcriptionally distinct haplotypes at this gene amongst inbred strains that are associated with resistance and susceptibility to *P. chabaudi*. In A/J, the corresponding mRNA carries a non-sense mutation that leads to premature termination of the encoded protein. This protein is involved in catabolic degradation of coenzyme A and production of cysteamine, a key metabolic precursor of glutathione in erythrocytes.

Conclusions and future perspectives

The impact of genomics on our ability to decipher the contribution of host genetics to susceptibility to infectious diseases has been tremendous, particularly in the mapping and identification of genes integral to the host response. Although the process of gene discovery using a forward genetic approach is relatively straightforward, in practice, the number of host resistance genes identified in this manner is relatively few. The number of QTLs being mapped has increased significantly in recent years, and as new technologies and methodologies become available, a corresponding increase in the pace of gene discovery underlying these mapped intervals is to be expected.

References

Burt RA, Baldwin TM, Marshall VM, Foote SJ 1999 Temporal expression of an H2-linked locus in host response to mouse malaria. Immunogenetics 50:278–285

Casanova JL, Abel L 2004 The human model: a genetic dissection of immunity to infection in natural conditions. Nat Rev Immunol 4:55–66

Cooke GS, Hill AV 2001 Genetics of susceptibility to human infectious diseases. Nat Rev Genet 2:967–977

Cuellar-Mata P, Jabado N, Liu J, Finlay BB, Gros P, Grinstein S 2002 Nramp1 modifies the fusion of *Salmonella typhimurium* containing vacuoles with cellular endomembranes in macrophages. J Biol Chem 277:2258–2265

Diez E, Lee SH, Gauthier S et al 2003 Birc1e is the gene within the Lgn1 locus associated with resistance to Legionella pneumophila. Nat Genet 33:55–60

Foote SJ, Burt RA, Baldwin TM et al 1997 Mouse loci for malaria-induced mortality and the control of parasitaemia. Nat Genet 17:380–381

Forbes JR, Gros P 2001 Divalent metal transport by Nramp proteins at the interface of host: parasite interactions. Trends Microbiol 9:397–403

Fortier A, Diez E, Gros P 2005 Naip5/Birc1e and susceptibility to Legionella pneumophila. Trends Microbiol 13:328–335

Fortin A, Belouchi A, Tam MF et al 1997 Genetic control of blood parasitaemia in mouse malaria maps to chromosome 8. Nat Genet 17:382–383

Fortin A, Diez E, Rochefort D et al 2001a Recombinant congenic strains derived from A/J and C57BL/6J: A tool for the genetic dissection of complex traits. Genomics 74: 21–35

Fortin A, Cardon LR, Tam M, Skamene E, Stevenson MM, Gros P 2001b Identification of a new malaria susceptibility locus (Char4) in recombinant congenic strains of mice. Proc Natl Acad Sci USA 98:10793–10798

Fortin A, Cardon LR, Stevenson MM, Gros P 2002 Complex genetic control of susceptibility to malaria in mice. Genes Immun 3:777–786

Fortin A, Diez E, Henderson J, Mogil JS, Gros P, Skamene E 2006 The AcB/Bca recombinant congenic strains of mice: strategies for phenotype dissection, mapping and cloning of quantitative trait genes. In: Decoding the genomic control of immune reactions. Wiley, Chichester (Novartis Found Symp 281) p 141–155

Frehel C, Canonne-Hergaux F, Gros P, de Chastellier C 2002 Effect of *Nramp1* on bacterial replication and on maturation of *Mycobacterium avium*-containing phagosomes in bone marrow-derived mouse macrophages. Cell Microbiol 4:541–556

Growney JD, Dietrich WF 2000 High resolution genetic and physical map of the *Lgn1* interval in C57BL/6J implicates Naip2 or Naip5 in *Legionella pneumophila* pathogenesis. Genome Res 10:1158–1171

Hernandez-Valladares M, Naessens J, Gibson JP et al 2004 Confirmation and dissection of QTL controlling resistance to malaria in mice. Mammal Gen 15:390–398

Hill AV 1998 The immunogenetics of human infectious disease Ann Rev Immunol 16: 593–617

Inohara O, Chamaillard M, McDonald C, Nunez G 2005 NOD-LRR proteins: role in host-microbial interactions and inflammatory disease. Annu Rev Biochem 74:355–383

Lam-Yuk-Tseung S, Gros P 2003 Genetic control of susceptibility to bacterial infections in mouse models. Cell Microbiol 5:299–313

Lee S-H, Girard S, Macina D et al 2001 Susceptibility to cytomegalovirus infection is associated with deletion of the natural killer cell lectin-like receptor *ly49h* gene. Nat Genet 28:42–45

Lee SH, Dimock K, Gray DA et al 2003a Maneuvering for advantage: the genetics of mouse susceptibility to virus infection. Trends Genet 19:447–457

Lee SH, Zafer A, de Repentigny Y et al 2003b Transgenic expression of the activating natural killer receptor Ly49H confers resistance to cytomegalovirus in genetically susceptible mice. J Exp Med 197:515–526

Li C, Seixas E, Langhorne J 2001 Rodent malarias: The mouse as a model for understanding immune response and pathology induced by erythrocytic stages of the parasite. Med Microbiol Immunol 189:115–126

Min-Oo G, Gros P 2005 Erythrocyte variants and the nature of their malaria protective effect. Cell Microbiol 7:753–763

Min-Oo G, Fortin A, Tam MF, Nantel A, Stevenson MM, Gros P 2003 Pyruvate kinase deficiency in mice protects against malaria. Nat Genet 35:357–362

Poon A, Schurr E 2004 The NRAMP genes and human susceptibility to common diseases. In: Cellier M, Gros P (eds) The Nramp Family. Kluwer Academic/Plenum Publishers, New York, NY p 29–43

Skamene E, Gros P, Forget A, Kongshavn PAL, St-Charles C, Taylor BA 1982 Genetic regulation of resistance to intracellular pathogens. Nature 297:506–509

Skamene E, Schurr E, Gros P 1998 Infection genomics: Nramp1 as a major determinant of natural resistance to intracellular infections. Ann Rev Med 49:275–287

Stevenson MM, Riley EM 2004 Innate immunity to malaria. Nature Rev Immunol 4: 169–180

Webb JR, Lee SH, Vidal SM 2002 Genetic control of innate immune responses against cytomegalovirus: MCMV meets its match. Genes Immun 3:250–262

Zaharik ML, Vallance BA, Puente JL, Gros P, Finlay BB 2002 Host-pathogen interactions: Murine host resistance factor Nramp1 upregulates *Salmonella enterica* serovar Typhimurium SPI2 virulence genes. Proc Natl Acad Sci USA 99:15705–15710

Zamboni DS, Kobayashi KS, Kohlsdorf T et al 2006 The Bircle cytosolic pattern-recognition receptor contributes to the detection and control of *Legionella pneumophila* infection. Nat Immunol 7:318–325

DISCUSSION

Hume: It is something of a *tour de force* to complement by BAC transgenics. What stops you from using retroviral complementation in bone marrow cells as a short-cut, especially if you have a presumptive myeloid phenotype?

Gros: This was something that was attempted by Bill Dietrich's lab in Boston. They spent a bit of time trying to do complementation in bone marrow macrophages without much success. In one paper he had some partial complementation, but clearly this was not a tool that would be very helpful. This is complicated by the fact that there are differences between our labs in the phenotypic evaluation of the LGN1 effect. In our lab we use the old protocol based on thioglycollate-elicited peritoneal macrophages. This is not a very nice population but this is what works for us. Bill uses bone marrow-derived macrophages but uses a different strain of *Legionella* to see his phenotype. The bottom line is that in the end, what worked for us was BAC transgenics. This also worked for Bill Dietrich. He also had some success with using RNAi as a method for suppressing *Naip5* expression in bone-marrow-derived macrophages, and having a partial effect on the phenotype. One has to think carefully before embarking on this because it is a long haul, but for complex regions I think that BAC transgenics are helpful. Sylvia Vidal has also done this for other complex regions such as the CMV1 on the LY49 cluster (Lee et al 2003). The added benefit in her studies was that she also had a BAC that rearranged when passed through the germline. This lost a piece, and this loss coincided with the loss of phenotype. It is an additional tool with which you can study large chunks of DNA. As opposed to YAC clones that can transfer much larger pieces of DNA but are inherently unstable, the BAC clones tend to be pretty stable.

Hume: I have a related question about the *Vanin* phenotype. What makes you assume that this is an erythroid phenotype rather than a myeloid one? And what is the explanation for the apparent pleiotropic effect on expression of *Vanin1*?

Gros: We have an answer to the second question. There is a cataclysmic event in the promoter region of the *Vanin* cluster that is specific for A/J. We have characterized this. We think that this probably impairs the overall transcriptional activity of the cluster to a significant degree. With respect to the first point, we have no proof whatsoever that this is an erythroid phenotype.

Hume: I have an interest in them as myeloid-specific genes.

Gros: We have the *Vanin1* knockout in the lab now. It would be nice to do the *Vanin3* knockout because this is also expressed in the myeloid compartment. Again, this is very new. We have no bias. We like to have a model up there because there is a connection to glutathione, but we haven't even done simple things such as knocking down GSH levels in red cells and looking at the effect.

Foote: Do you think there is a specific reason that the PK animals are resistant, or do you think that just altering the red cells is enough to make an animal resistant to malaria?

Gros: We thought it would be nice if there was accumulation of the metabolic intermediates upstream of PK that would make it less hospitable for *Plasmodium* to replicate. We have done experiments to look at the red cell turnover in those animals. My feeling is that you can explain most of the phenotype by faster red cell turnover in the PK-deficient animals. The independent CBA PK-deficient allele is a more severe PK deficiency than the ACB55 PK55 mutant, and in this strain there is an even faster red cell turnover rate, along with a stronger protection against malaria. A similar but preliminary result was obtained by looking at replication of *P. falciparum* in primary erythrocyte cultures from PK-deficient humans. It doesn't look like it is the ability of the parasite to switch forms once it is inside the red cell that is altered; the resistance is probably explained by a different half life and turnover rate. We would like to have PK as a therapeutic target of some kind, but we think the results can be explained mechanically.

Maizels: There are some species of malaria that seem to prefer reticulocytes as host cells. Is it possible that you have a mutation that decreases susceptibility to one species but increases it to another?

Gros: That's a great question. We are aware of this and would like to do the experiments. If it works out this way it would be very nice.

Goodnow: How do you explain the recrudescence of parasitaemia in the IRF8 mutants, and the long-term progression of *Mycobacterium bovis* infection? Could this be due to some sort of exhaustion of helper T cells? This gets to a much more general issue of why most of us don't develop a full sterilizing immunity to tuberculosis (TB) or ever get rid of malaria parasites.

Gros: One of the things that I was expecting from this talk is that there would be some feedback from immunologists. As you say, the data are striking. There is a parallel effect on *Salmonella*, where we see a clear effect of loss of function of IRF8 on susceptibility to *Salmonella*. With mycobacteria we have done pretty straightforward histology on granulomas. Clearly, the 'granulomas' that form in those BXH2 mice are not very good. There are fewer of them and they are not as well organized. It looks as if there is defective containment of infection at the infection site in those animals. In malaria, I have no idea, but it is a phenotype very similar to the IL12 knockout, with recrudescent waves of parasitaemia. You clearly need an isotype switch or something else with respect to innate immunity.

The parasitaemia goes right down, almost to zero: mice are able to clear most of it but not all, because it comes back, presumably through some kind of antigenic switch in the parasite. The ability to respond to this properly is IL12 dependent. But I have no idea which cells are involved.

Casanova: It might be a useful experiment to assess both the microbe-produced IL12 production and also the $CD4^+$ T cell-dependent IL12 production. The first one most likely will be affected by your mutation. The $CD4^+$ T cell-dependent pathway may or may not be influenced by IFNγ and ICSBP. This could be a straightforward experiment for detecting the impact of your genetic lesion on the two pathways of production.

Britton: Does it have a similar effect on TB infection in mice?

Gros: We will tell you in a few weeks. The ICSBP mutation also affects the myeloid compartment, so these mice develop a CML-type phenotype. The absence of ICSBP shuts down the mononuclear pathway and causes an expansion of the granulocyte pathway. These mice have a CML-like disease in which they are deficient in the monocyte kind of progenitor from the same stem cells. They die early of this leukaemia. It is a complex story because they also have a replication-competent retrovirus that specifically replicates inside that expanding progenitor. They are hard to breed, and if you want to look at TB they have to live long enough that you can phenotype them.

Britton: What happens to the T cell response later at the 8 week time point? Is any measure of the T cell response abnormal?

Gros: This is something that we are doing now. We are doing cytokine arrays and seeing what kind of pleiotropic effects the loss of ICSBP or of the myeloid mononuclear compartment has on the overall cytokine profile.

Goodnow: Gabrielle Belz, in terms of your work looking at T cells and the generation of memory versus exhaustion, is there any connection between IFNγ, IRF8 and different dendritic cell (DC) subsets, and what might preserve versus exhaust your T cell pool?

Belz: We have a model with Brendan Crabb of transgenic malaria. We found that the $CD8^+$ DCs are the main ones presenting the antigen. In the IRF8 knockouts, the $CD8^+$ DCs are almost completely ablated. There is also a shut-down of the ability of the T cells to respond to any cells presenting antigens a short while after they are first activated. The mechanism isn't known. There is a complete crippling of the CD8 arm.

Turner: There is also an effect in the IFNγ knockouts and IFNγ receptor 1 knockouts, although the literature is a little controversial. John Harty's group has shown (Badovinac et al 2000) that lack of IFNγ in certain viral and bacterial infections results in normal expansion of CD8 numbers, but there is a defect in the contraction phase and formation of memory. Lindsey Whitton's group (Whitmire et al 2005) used IFNγ receptor knockouts and found that there is a defect in the expansion phase. There is no clear explanation for why if you knock out the cytokine there is such a different phenotype versus the actual signalling events.

References

Badovinac VP, Tvinnereim AR, Harty JT 2000 Regulation of antigen-specific $CD8^+$ T cell homeostasis by perforin and interferon-gamma. Science 290:1354–1358

Lee SH, Zafer A, de Repentigny Y et al 2003 Transgenic expression of the activating natural killer receptor Ly49H confers resistance to cytomegalovirus in genetically susceptible mice. J Exp Med 197:515–526

Whitmire JK, Tan JT, Whitton JL 2005 Interferon-gamma acts directly on $CD8^+$ T cells to increase their abundance during virus infection. J Exp Med 201:1053–1059

Mycobacterium tuberculosis and its ability to resist immunity

Douglas Young*† and Anne O'Garra†

**Imperial College London, South Kensington Campus, London SW7 2AZ and †National Institute for Medical Research, Mill Hill, London NW7 1AA, UK*

Abstract. Resistance to tuberculosis involves a balance between the immune activation required to restrict the infection and the immune regulation required to prevent collateral damage to surrounding tissues. We explore here the hypothesis that genetic diversity of *Mycobacterium tuberculosis* provides an opportunity to tilt this balance in favour of the pathogen through variations in innate immune signalling.

2007 Decoding the genomic control of immune reactions. Wiley, Chichester (Novartis Foundation Symposium 281) p 169–180

Nine million people develop active tuberculosis (TB) every year, resulting in two million deaths. The pathogen, *Mycobacterium tuberculosis*, is transmitted by aerosol from the lungs of individuals with active disease, resulting in an annual rate of infection of 1–2% in populations living in endemic areas. Only a small fraction of infected individuals, around 5–10%, go on to develop active TB. For the vast majority of individuals the natural immune response is sufficient to control the infection, though probably often leaving residual viable bacteria in the form of a latent infection with the potential for subsequent reactivation. Impairment of the immune response—as a result of coinfection with HIV for example—dramatically increases the risk of progression to active disease. A major challenge for TB disease control programmes is to engineer the reverse process of reducing progression to disease by closing the 10% gap left by natural immunity (Young & Dye 2006).

Studies of human genetic diversity clearly demonstrate a role for the host genome in control of immunity to TB. Compelling evidence comes from comparison of disease in monozygotic and heterozygotic twins (Comstock 1978), and from the identification of individuals with Mendelian susceptibility to mycobacterial disease (Alcais et al 2005). Several polymorphisms have been linked to TB: a subset of these is shown in Table 1. Genes implicated in Mendelian susceptibility are involved in the central process of the interleukin (IL)12/interferon (IFN)γ activation

TABLE 1 Examples of human gene polymorphisms associated with differential susceptibility to mycobacterial disease

Gene	*Study population*	*References*
HLA-DR2, HLA-DQ1	Pulmonary TB effects differ between populations	Marquet & Schurr 2001
NRAMP1 (SLC11A1)	Different effect in pulmonary *versus* paediatric TB	Malik et al 2005
vitamin D receptor	Pulmonary TB	Bellamy et al 1999
mannose binding lectin	Different effect in pulmonary *versus* paediatric TB	Selvaraj et al 1999, Hoal-Van Helden et al 1999
IFNγ receptor	Hypersusceptible children	Alcais et al 2005
IL12	Hypersusceptible children	Alcais et al 2005
IL12 receptor	Hypersusceptible children	Alcais et al 2005
STAT1	Hypersusceptible children	Alcais et al 2005
MCP-1	Pulmonary TB	Flores-Villanueva et al 2005
DC-SIGN	Pulmonary TB	Barreiro et al 2006

pathway and have a profound impact on the immune response to mycobacteria. Other polymorphisms exert a more subtle influence on the risk of disease, suggesting a role in fine-tuning of the immune response. Different selective pressures may be operating on different sets of genes. TB commonly occurs as disseminated, often fatal disease following initial infection in children, or as pulmonary disease associated with reinfection or reactivation in adults. Genes that are differentially involved in disease onset before and after reproduction may have been subject to different forms of selection.

Is the immune response to mycobacteria also under the control of the microbial genome? Viewed from the perspective of *M. tuberculosis*, the 10% failure rate associated with natural immunity fully supports transmission of infection and maintenance of a healthy microbial population. Perhaps this represents a window generated by active immune subversion by the pathogen? Or, from a more interactive viewpoint, pathogen variants may have been selected to exploit niches associated with variations in the immune response. To parallel the human studies, we can look for evidence of this in the genetic diversity of *M. tuberculosis*.

Genetic diversity of *M. tuberculosis*

Classical microbiologists observed that clinical isolates of *M. tuberculosis* displayed phenotypic diversity in laboratory culture and in animal models, but molecular genetic analysis has begun to explore this phenomenon only in the last few years. Strains responsible for human TB belong to the '*M. tuberculosis* complex', which also includes strains that cause TB in a range of animal species (*M. bovis*, *M. microti* and *M. pinnipedii* cause disease in cattle, voles and seals, for example). Members of the *M. tuberculosis* complex are 99.95% identical at the nucleotide level (Garnier et al 2003)—comparable to the 0.1% overall diversity present in humans—and probably represent clonal expansion from some form of genetic bottleneck within the last 50 000 years. One attractive hypothesis is that the current strains emerged from an ancestral family of pathogens referred to as '*Mycobacterium prototuberculosis*'. It has been suggested that strains with a characteristic smooth colony morphology that are associated with human disease in the Horn of Africa represent extant members of this progenitor family (Gutierrez et al 2005).

Large sequence polymorphisms provide a powerful tool for genetic mapping of the *M. tuberculosis* complex. These represent deletion events resulting in loss of 1–20 kb segments of the genome and, since there is no evidence that *M. tuberculosis* acquires genes by horizontal transfer, it can be inferred that strains lacking these sequences have evolved from strains in which these regions are intact. Using a set of deletions and selected single nucleotide polymorphisms (SNPs), the *M. tuberculosis* complex can be presented as a series of variants that have adapted to different mammalian hosts (Brosch et al 2002). At the level of experimental infection, predilection for a particular host is seen as relative rather than absolute, but subtle changes in host–pathogen balance may be crucial in the overall dynamics of disease transmission in the natural environment. Human isolates of *M. tuberculosis* readily cause infection in mice and guinea-pigs, for example, but productive transmission is seen only in humans and elephants. *M. bovis* isolates from cattle are able to cause human disease, but at a population level are significantly less successful than *M. tuberculosis*.

Human *M. tuberculosis* lineages can be subdivided on the basis of further large sequence polymorphisms, SNPs and other small deletions (Gagneux et al 2006, Filliol et al 2006, 2003). Movement of insertion elements and changes in copy number of tandem repeats provide further fine detail for epidemiological mapping of local outbreaks and transmission chains (van Soolingen 2001). A global phylogeography of *M. tuberculosis* is beginning to emerge from such studies, with different lineages found to predominate in different parts of the world. Interestingly, the association of different lineages with particular human populations is maintained during migration. *M. tuberculosis* lineage is preserved amongst ethnic groups settled in San Francisco, for example (Hirsh et al 2004, Gagneux et al 2006), and a human migration model has been used to map the distribution of members of the predominant Beijing lineage in Asia and America (Mokrousov et al 2005). While cultural and social behaviour patterns will have an impact on disease transmission, it might have been anticipated that, as an airborne pathogen, *M. tuberculosis* would rapidly equilibrate during the intermixing of populations. Rapid dissemination has been reported in some settings, particularly in the case of members of the

Beijing lineage in the context of TB outbreaks and drug resistance (Glynn et al 2002, Bifani et al 2002). This has prompted suggestions that Beijing strains represent a hypervirulent phenotype, perhaps with a selective advantage in BCG-vaccinated populations (van Soolingen et al 1995).

While the absence of comprehensive global data makes it hard to exclude the effects of selective sampling in these studies, a model emerging from population biology would envisage parasitism by '*M. prototuberculosis*' extending perhaps throughout the whole of human evolution, followed by the recent clonal expansion of the *M. tuberculosis* complex, and ongoing selection of variants with predilection for species-specific niches. It is possible that genetic and epidemiological differences amongst human populations provide multiple specialised niches that favour different pathogen variants. Is there any link between genotypic diversity and biological phenotype that might support this model?

Phenotypic diversity and the innate immune response

Sequenced isolates of *M. tuberculosis* and *M. bovis* differ by 64 genes (all lost from *M. bovis*) and of the order of 1000 SNPs (Garnier et al 2003). This provides a reasonably manageable palette to begin to search for species-specifying polymorphisms. While the lineage-defining deletions are an attractive starting point, there is a good chance that these are functionally neutral markers that might 'hitch-hike' with biologically relevant changes elsewhere on the genome. To date, the best example of a genotype–phenotype association has come from study of the hypervirulent phenotype of the Beijing lineage (Reed et al 2004). This led to identification of a phenolic glycolipid (PGL) as a key mediator of the rapid lethal phenotype characteristic of a Beijing isolate in a mouse model. Production of PGL is dependent on a polyketide synthase gene (*pks15/1*). The majority of *M. tuberculosis* lineages have inherited a mutation in which the *pks15/1* gene has been inactivated by a 7 bp deletion; the Beijing lineage is unusual in having retained the ancestral functional genotype (Constant et al 2002). Experimental inactivation of *pks15/1* in the Beijing isolate resulted in loss of PGL and reversal of the lethal phenotype in mice (Reed et al 2004).

This study also identified a mechanism for the effect of PGL (Reed et al 2004). Incubation of bone marrow macrophages with purified PGL, or infection with PGL-expressing *M. tuberculosis*, suppressed the innate immune response measured by production of TNFα and IL6. By suppressing initial innate recognition, the Beijing strain delays induction of clearance by the adaptive immune response, resulting in a more aggressive disease. In these studies, PGL acts as a classical virulence factor enhancing pathogenesis of the Beijing strain. However, the fact that the majority of the successful strains of *M. tuberculosis* have dispensed with PGL synthesis suggests a more complex provenance. It may be that suppression

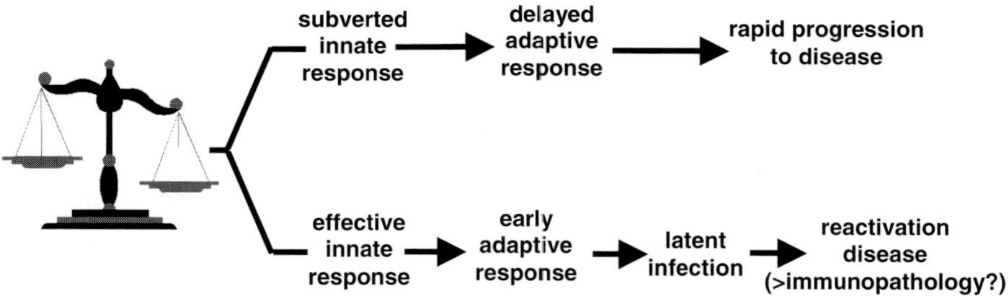

FIG. 1. A pivotal role for innate immune signalling. Clinical isolates of *M. tuberculosis* differ with respect to the innate immune response. Strains that suppress innate immune signalling (members of the Beijing lineage, for example) may achieve an advantage by delaying onset of adaptive immunity. Strains that elicit a strong immune response may achieve some other advantage; increased immune pathology may enhance transmission, for example.

of the innate immune response also has a downside for *M. tuberculosis*. Suppression may have a negative impact on the immune pathology required for lung damage and efficient transmission, for example. It is attractive to propose a model in which *M. tuberculosis* would find advantage at either end of an innate immunogenicity spectrum (Fig. 1). Different strains might then position themselves along this spectrum by pushing the appropriate series of buttons involved in innate immune recognition; this would translate into different fitness profiles appropriate for success in different epidemiological settings and host genetic backgrounds. A prediction from this model is that phenotypic diversity would be associated with changes in the repertoire of ligands involved in triggering innate immune recognition.

Modelling innate immunity

The outer envelope of mycobacteria is rich in the classes of lipids, glycolipids, lipoproteins, glycoproteins and polysaccharides that typically provide ligands for pattern recognition receptors of the innate immune system. The fine detail of structural modifications to some of these macromolecules—protein glycosylation (Herrmann et al 1996), and mannose capping of the arabinose ends of lipoarabinomannan (Khoo et al 2001), for example—may have a critical influence on recognition processes, moderating differential signalling pathways through Toll-like and C-type lectin receptors (Kaufmann & Schaible 2003). In addition to the concentrated set of ligands that the innate immune cell will confront on the surface of a bacterium, further isolated stimuli will be provided by ligands that are shed from extracellular organisms or secreted from infected (Beatty et al 2000) and apoptotic cells (Winau et al 2006). All of these factors will influence the decision making process by which the innate immune system sets the environment for the subsequent adaptive response. Many of the same mycobacterial ligands are also recognised as antigens by T cell subsets restricted by CD1 molecules and carrying a $\gamma\delta$ receptor. Given such complexity, it is easy to imagine that subtle qualitative or quantitative changes affecting mycobacterial cell wall components might have an impact on the immune response, but it is hard to envisage a simple model that would allow any form of predictive analysis.

The task facing the innate immune system is a generic problem of establishing a network that links a complex series of potential outputs to a diverse range of inputs. A common solution to creating a robust system to address this problem is in the form of a 'bow-tie' architecture, consisting of highly conserved efficient core processes flanked by flexible external domains (Fig. 2) (Csete & Doyle 2004). Such systems demonstrate 'highly optimized tolerance', being robust against specific perturbations although fragile against unexpected perturbations (Carlson & Doyle 2002). For biological systems, this type of network has been explored particularly for modelling of metabolism, in which a wide variety of nutritional inputs are converted to a wide range of synthesised products by way of a limited set of key intermediates. Kitano and Oda have recently applied this approach to model the immune system (Kitano & Oda 2006). They envisage a nested bow-tie structure with conserved core processes represented by MyD88 signalling for innate immunity, and TCR recognition of peptide–MHC for the adaptive response. It is attractive to think that network analysis at this level may begin to provide a framework for dealing with the complexity of innate immune recognition.

Kitano and Oda went on to consider the potential importance of gut flora in establishing and maintaining this network (Kitano & Oda 2006). There is accumulating evidence that the host genetic background has an important role

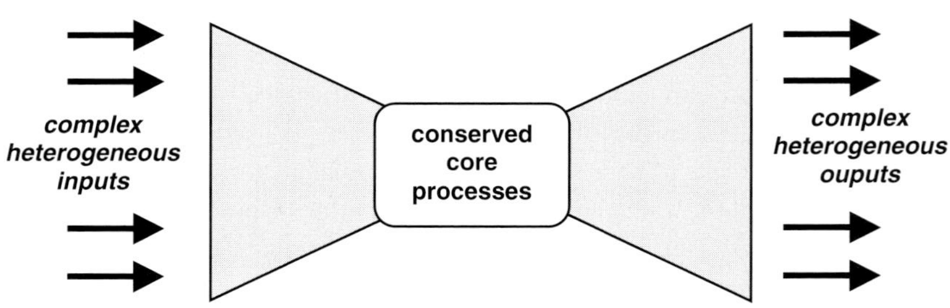

FIG. 2. A 'bow-tie' system in which a highly conserved core process is flanked by flexible external domains provides a robust solution for dealing with a high degree of complexity. This has been used to model biological networks, including metabolism and, more recently, the immune system. (Adapted from Csete & Doyle 2004.)

in determining the composition of gut microflora, and that this in turn has an important impact on the way that different individuals respond to drugs (Nicholson et al 2005). A prediction from the network analysis is that genetics and environmental exposure may have a similar impact on the way that different individuals respond to vaccines and indeed to infection in general. In the context of TB, this type of thinking may help rationalise concepts related to the role of environmental mycobacteria in modulating responses to BCG (Black et al 2002) as well as the potential impact of worm and other parasitic infections (Elias et al 2005).

A challenge for systems biology is to find ways in which knowledge of environmental influences can be integrated with information about host and pathogen genetic variation in order to predict susceptibility to TB and to promote rational design of vaccines and other means of enhancing resistance.

References

Alcais A, Fieschi C, Abel L, Casanova JL 2005 Tuberculosis in children and adults: two distinct genetic diseases. J Exp Med 202:1617–1621

Barreiro LB, Neyrolles O, Babb CL et al 2006 Promoter variation in the DC-SIGN-encoding gene CD209 is associated with tuberculosis. PLoS Med 3:e20

Beatty WL, Rhoades ER, Ullrich HJ et al 2000 Trafficking and release of mycobacterial lipids from infected macrophages. Traffic 1:235–247

Bellamy R, Ruwende C, Corrah T et al 1999 Tuberculosis and chronic hepatitis B virus infection in Africans and variation in the vitamin D receptor gene. J Infect Dis 179:721–724

Bifani PJ, Mathema B, Kurepina NE, Kreiswirth BN 2002 Global dissemination of the Mycobacterium tuberculosis W-Beijing family strains. Trends Microbiol 10:45–52

Black GF, Weir RE, Floyd S et al 2002 BCG-induced increase in interferon-gamma response to mycobacterial antigens and efficacy of BCG vaccination in Malawi and the UK: two randomised controlled studies. Lancet 359:1393–1401

Brosch R, Gordon SV, Marmiesse M et al 2002 A new evolutionary scenario for the Mycobacterium tuberculosis complex. Proc Natl Acad Sci USA 99:3684–3689

Carlson JM, Doyle J 2002 Complexity and robustness. Proc Natl Acad Sci USA 99 Suppl 1:2538–2545

Comstock GW 1978 Tuberculosis in twins: a re-analysis of the Prophit survey. Am Rev Respir Dis 117:621–624

Constant P, Perez E, Malaga W et al 2002 Role of the pks15/1 gene in the biosynthesis of phenolglycolipids in the Mycobacterium tuberculosis complex. Evidence that all strains synthesize glycosylated p-hydroxybenzoic methly esters and that strains devoid of phenolglycolipids harbor a frameshift mutation in the pks15/1 gene. J Biol Chem 277:38148–38158

Csete M, Doyle J 2004 Bow ties, metabolism and disease. Trends Biotechnol 22:446–450

Elias D, Akuffo H, Pawlowski A et al 2005 Schistosoma mansoni infection reduces the protective efficacy of BCG vaccination against virulent Mycobacterium tuberculosis. Vaccine 23:1326–1334

Filliol I, Driscoll JR, van Soolingen D et al 2003 Snapshot of moving and expanding clones of Mycobacterium tuberculosis and their global distribution assessed by spoligotyping in an international study. J Clin Microbiol 41:1963–1970

Filliol I, Motiwala AS, Cavatore M et al 2006 Global phylogeny of Mycobacterium tuberculosis based on single nucleotide polymorphism (SNP) analysis: insights into tuberculosis evolution, phylogenetic accuracy of other DNA fingerprinting systems, and recommendations for a minimal standard SNP set. J Bacteriol 188:759–772

Flores-Villanueva PO, Ruiz-Morales JA, Song CH et al 2005 A functional promoter polymorphism in monocyte chemoattractant protein-1 is associated with increased susceptibility to pulmonary tuberculosis. J Exp Med 202:1649–1658

Gagneux S, Deriemer K, Van T et al 2006 Variable host-pathogen compatibility in Mycobacterium tuberculosis. Proc Natl Acad Sci USA

Garnier T, Eiglmeier K, Camus JC et al 2003 The complete genome sequence of Mycobacterium bovis. Proc Natl Acad Sci USA 100:7877–7882

Glynn JR, Whiteley J, Bifani PJ, Kremer K, van Soolingen D 2002 Worldwide occurrence of Beijing/W strains of Mycobacterium tuberculosis: a systematic review. Emerg Infect Dis 8:843–849

Gutierrez MC, Brisse S, Brosch R et al 2005 Ancient origin and gene mosaicism of the progenitor of Mycobacterium tuberculosis. PLoS Pathog 1:e5

Herrmann JL, O'Gaora P, Gallagher A, Thole JE, Young DB 1996 Bacterial glycoproteins: a link between glycosylation and proteolytic cleavage of a 19 kDa antigen from Mycobacterium tuberculosis. EMBO J 15:3547–3554

Hirsh AE, Tsolaki AG, DeRiemer K, Feldman MW, Small PM 2004 Stable association between strains of Mycobacterium tuberculosis and their human host populations. Proc Natl Acad Sci USA 101:4871–4876

Hoal-Van Helden EG, Epstein J, Victor TC et al 1999 Mannose-binding protein B allele confers protection against tuberculous meningitis. Pediatr Res 45:459–464

Kaufmann SH, Schaible UE 2003 A dangerous liaison between two major killers: Mycobacterium tuberculosis and HIV target dendritic cells through DC-SIGN. J Exp Med 197:1–5

Khoo KH, Tang JB, Chatterjee D 2001 Variation in mannose-capped terminal arabinan motifs of lipoarabinomannans from clinical isolates of Mycobacterium tuberculosis and Mycobacterium avium complex. J Biol Chem 276:3863–3871

Kitano H, Oda K 2006 Robustness trade-offs and host-microbial symbiosis in the immune system. Mol Syst Biol 2:2006.0022

Malik S, Abel L, Tooker H et al 2005 Alleles of the NRAMP1 gene are risk factors for pediatric tuberculosis disease. Proc Natl Acad Sci USA 102:12183–12188

Marquet S, Schurr E 2001 Genetics of susceptibility to infectious diseases: tuberculosis and leprosy as examples. Drug Metab Dispos 29:479–483

Mokrousov I, Ly HM, Otten T et al 2005 Origin and primary dispersal of the Mycobacterium tuberculosis Beijing genotype: clues from human phylogeography. Genome Res 15: 1357–1364

Nicholson JK, Holmes E, Wilson ID 2005 Gut microorganisms, mammalian metabolism and personalized health care. Nat Rev Microbiol 3:431–438

Reed MB, Domenech P, Manca C et al 2004 A glycolipid of hypervirulent tuberculosis strains that inhibits the innate immune response. Nature 431:84–87

Selvaraj P, Narayanan PR, Reetha AM 1999 Association of functional mutant homozygotes of the mannose binding protein gene with susceptibility to pulmonary tuberculosis in India. Tuber Lung Dis 79:221–227

van Soolingen D 2001 Molecular epidemiology of tuberculosis and other mycobacterial infections: main methodologies and achievements. J Intern Med 249:1–26

van Soolingen D, Qian L, de Haas PE et al 1995 Predominance of a single genotype of Mycobacterium tuberculosis in countries of east Asia. J Clin Microbiol 33:3234–3238

Winau F, Weber S, Sad S et al 2006 Apoptotic vesicles crossprime CD8 T cells and protect against tuberculosis. Immunity 24:105–117

Young D, Dye C 2006 The development and impact of tuberculosis vaccines. Cell 124: 683–687

DISCUSSION

Mackay: I'd like to pick up the business about the environment. My understanding is that prior exposure to environmental mycobacteria has a big effect on whether vaccination to TB works. Can you expand on this for me?

Young: Where the environmental mycobacteria story has come in is in terms of trying to explain BCG vaccination. This works well in the UK but doesn't work in the tropics. The inference is that it tends to work in places where there are cooler climates and less environmental mycobacteria. There are two explanations. First, when you are exposed to environmental mycobacteria you set up some level of anti-mycobacterial immunity. In order for BCG to be effective it needs to replicate, but because there is some kind of baseline mycobacterial immunity, the vaccine doesn't take. The other explanation is that some of the environmental mycobacteria are able to subvert the immune response. So rather than getting a pure Th1, which is what we think we would want for TB protection, somehow environmental mycobacteria are skewing some Th2 response. Overall, there are various bits of evidence that suggest that primary exposure to mycobacteria in the environment has some kind of impact on our response to vaccination and TB. I don't know how we feed this into our model in mechanistic terms.

Britton: The accelerated lethality still occurs at day 200, which is a bit daunting for any experimental model looking at genetic control. Identifying the factors that control or regulate this in the host is difficult, but this is where considerable variation occurs among individuals. What are the genetic factors in a person without major defects in cellular immunity that cause delayed reactivation of infection and lung cavitation?

Young: A more optimistic way to spin this is that if you are convinced of the importance of the innate immune response, you may say what is really important is the early events in the initial innate immune response. What is seen at day 200 could be a downstream consequence of something that happened on day 7. This would reduce the mouse costs tremendously.

Britton: Factors which regulate the capacity for apoptosis in infected cells may have influence on the nature of the T cell response early in the infection. This is evident later by reactivation of infection.

Goodnow: Why do you assume that this is acting only on the innate response? Could phenolic glycolipids be having an effect on T cell differentiation, and memory versus exhaustion issues?

Young: There is an effect on T cells, but it can explained by the reduced initial innate immune signalling.

Goodnow: If you test the same phenolic glycolipids, do they have direct effects on lymphocytes themselves?

Young: Not that I am aware of. No one has shown anything.

Karupiah: I want to ask about the interaction of the pathogen with the host. There are different types of pathogen, and the host will shape its response accordingly. The way one pathogen might help shift the immune response might not be good for another pathogen. It is a lot more complicated than looking at just a single pathogen–host interaction, as I'm sure you would agree.

Young: I would include that in the environment issue. Gut microflora may be just as important as pathogens. We need some way of embracing that complexity.

Turner: Is there any hint from the genes that are missing from *M. bovis* compared to *M. tuberculosis* that might help explain the species barrier?

Young: Nothing very clear so far. Quite a lot of effort will go into explaining what the difference is because the answer must be somewhere within the 2000 SNPs and 80 gene deletions. There is nothing compelling that says that deletions are the functional difference. My impression is growing that those deletions are just neutral markers of lineages. I don't think these are the phenotypically defining lesions. I suspect they are genes surplus to requirements.

Turner: As you go down those lineages you showed, is there an increased amount of deletion, or does it differ from one species to another?

Young: You can map out a progression; *M. bovis* has more deletions than *M. microti* which has more deletions than *M. tuberculosis*. But in human strains, the more we look the more deletions we find.

Foote: We are interested in the idea that there are strains that are associated with different ethnic groups. There also seems to be a fairly strong environmental influence. Is it not environmental influence if the gut flora is different in people who have culturally different diets?

Young: It will be difficult to disentangle. As well as environmental, there are social factors that will be important. In some circumstances strains are staying with particular groups; in others, such as the Beijing strain, they diffuse out in apparently similar social structures.

Hume: Warwick Britton raised the issue of these animals dying at 200 days instead of 300. Is the infection you are looking at solely a mycobacterial infection? Or is the progression precipitated by diminished resistance to hut flora? In the average animal house some kind of lung-active virus will be present. The relationship then is whether the mycobacterial products are actually suppressing the response to the secondary triggering agent, rather than the bacterium itself.

Young: I can't think of any evidence for this. I think it is unlikely.

Hume: It is the same as the exacerbation of inflammatory diseases. This is the paradigm of progression of Alzheimer's and chronic obstructive pulmonary disease

(COPD) where there are exacerbation events triggered by viral infections which lead to the progression of the disease.

Young: We think that the progression of the disease is due largely to the immune response. Eventually their lungs just gum up. The mycobacterial numbers have reached a plateau but there is progressive lung damage by infiltration of T cells.

Hume: Is that due to a response to the bug itself, or a secondary challenge?

Young: The endogenous retrovirus is hard to exclude. The others can be excluded by the fact that in different animal houses people get the same data with the same strains of mice.

Britton: There are notable examples of how one infection influences the responses to other pathogens, for example measles and TB. There are also intriguing data from west Africa where BCG immunization is associated with improved responses to other vaccines and reduced mortality from unrelated infectious diseases. This effect may not be seen in the north, but only in environments where there are multiple infectious threats occurring early on in life. This interaction is not readily analysed in animal models, but challenges us to develop relevant animal models to test these hypotheses.

Goodnow: There is an interesting example from Sewell-Wright's treatise, where he recounts this early work done on different guinea-pig strains on their capacity to resist TB infection. Initially the strains were characterized into varying degrees of resistance at the Walter Reed Army Hospital. Then when the investigator who did all this work moved to the better-resourced Rockefeller Institute, the susceptible strains were all resistant as well, so he was unable to the genetics. It turned out that the Walter Reed ration for the guinea-pigs was suboptimal, and in this suboptimal nutrition it was possible to see the effects of genetics on resistance to TB. When all the guinea-pigs were well fed this was gone. This is an issue for us.

Gros: From our experience, if we look at TB in a genetic framework in large crosses, we get some quantitative trait loci (QTLs) that are specific if time of survival is used as a quantitative trait. Then if we use pulmonary replication as our trait, we will also get QTLs for that. But we also get QTLs that come up with both replication at early time points and death. For those it would be difficult to infer a second infectious aetiology. But for those that are specific for one or the other, it is formally possible that external factors may contribute.

Morahan: I'll change tack and ask about the age-dependent switch between the generalized TB and the re-emergent TB. This has a parallel in type 1 diabetes, where there seems to be a major incidence before the age of 5 and then again around about the age of puberty. Do you have any idea about what it is about the changing nature of the immune response that causes this re-emergence phenomenon?

Young: I think it is dramatic, and it is an obvious question. It is not one that fits into the models of the immune response that I am aware of.

Britton: Also, there is the male–female difference with boys more likely to die from early childhood infections than girls.

Casanova: In a genome-wide scan previously reported, there was a possible candidate locus on the X-chromosome that might account for the sex ratio bias well documented in adults (but not in children).

Systems genetics: the next generation in genetics research?

Grant Morahan and Robert W. Williams*

*Centre for Diabetes Research, The Western Australian Institute for Medical Research, University of Western Australia, Medical Research Foundation Building, 50 Murray St, Perth, WA 6000, Australia and *Department of Anatomy and Neurobiology, Developmental Genetics, Department of Pediatrics, University of Tennessee Health Science Center, 855 Monroe Avenue, Memphis, TN 38163, USA*

This presentation was given at short notice to fill a gap in the meeting's scheduled program. The subject 'systems genetics' was chosen to alert participants to this new powerful technology which we see as the next generation in genetics research. It is fair to say that most scientists still see genetics as Mendel envisioned it: this gene codes for this product/protein/phenotype. It is increasingly clear that many of the major public health diseases we face—diabetes, cardiovascular diseases, cancers, etc.—although genetically based, do not follow this paradigm: understanding such non-Mendelian, complex genetic diseases with involvement of multiple genes and environmental factors is the challenge of what may be considered as the second generation genetics. Beyond this, the field we have dubbed systems genetics offers the opportunity to define interacting clusters and networks of genes within a tissue or cell population. This work has been pioneered by Rob Williams and colleagues (Chesler et al 2004, 2005, Li et al 2005) and taken up by other groups (Bystrykh et al 2005, Hubner et al 2005, Williams et al 2006). Formerly termed 'genetical genomics' we consider 'systems genetics' a more useful description. Here, we will describe some applications of this technology, using as an example the only immune-related tissue for which data is currently available: haemopoietic stem cells (Li et al 2005).

Systems genetics utilizes microarray technology, but can be considered as two quanta levels beyond conventional microarrays. The first quantum difference is that, instead of examining tissues from a single source, tissues from as many as 100 related strains are examined. The second level is that the transcriptomic data is integrated with the underlying genotypic data of each of the strains. In this way, it becomes possible to define clusters and networks of co-regulated transcripts, to correlate phenotype with genotype and to map genes which regulate the expression of other genes.

For this work to be possible, two resources are required. The first is a collection of related, genetically defined individuals that can be used as a source of tissues.

The 'systems' aspect of systems genetics implies the acquisition and correlation of a range of phenotypes; this is obviously much more difficult if one does not have an immortal mapping population. So, while it is possible to use F2, backcross or even heterogenous stock animals for this purpose, these are obviously limited as each animal has a unique genotype and can offer definition of only a limited number of phenotypes. Recombinant inbred (RI) mouse strains are ideal for this purpose. Each strain is derived from two progenitor strains and has a mosaic genome comprised of chromosome segments inherited from one of the two parental strains. Perhaps the most studied RI set is the BXD series, derived from C57BL/6 and DBA/2 strains (Taylor 1989). An original panel of 26 strains produced by Benjamin A. Taylor from the late 1970s (Taylor 1989) has since been supplemented with an additional 63 strains (Peirce et al 2004). Importantly, all mice within a strain are essentially genetically identical, so results may be accumulated from different tissues at different time points and under differing stimulation conditions; results may be confirmed subsequently and by other researchers. Each strain has inherited a different pattern of genes, so differences between strains can be mapped on the basis of allele sharing by similar strains. Both quantitative and qualitative traits may be mapped using RI strains (Taylor 1989).

The other resource required for this work is a new set of analytical tools. The *GeneNetwork* suite of programs, incorporating WebQTL (Chesler et al 2004, Wang et al 2003), is a very powerful resource for systems genetics analyses. As the name suggests, WebQTL is available online (*www.genenetwork.org*), facilitating access by researchers worldwide to a range of accumulated data.

We will illustrate some of the features of WebQTL below, As an example, we will look at expression in haemopoietic stem cells (HSC) of *Mlf1*, the gene encoding myeloid leukaemia factor (Yoneda-Kato et al 1996). The microarray data for this cell population were generated at the Genomics Institute of the Norvartis Research Foundations and at the University of Groningen as described by Bystrykh et al (2005). Figure 1 shows the variation in expression of *Mlf1* between the BXD strains. This variation, normally distributed across a 10-fold range, suggests that the level of *Mlf1* expression in HSC is under some genetic control. One of the WebQTL modules allows mapping of quantitative trait loci. Figure 1 also shows results of such QTL linkage mapping. A locus on chromosome 10 has genome-wide significance in linkage to modulation of *Mlf1* expression levels. (The *Mlf1* gene itself is on chromosome 3.)

We can now ask the question: as *Mlf1* expression rises or falls across the panel of strains, what other transcripts also rise (or fall) in the same strains? What genes may decrease in expression as *Mlf1* expression increases? In other words, can we correlate any other genes' expression with that of *Mlf1*? The results shown in Fig. 2 indicate that it is possible to define a cluster of apparently co-regulated genes. Some show an inverse correlation; the P values for the correlation of these genes are all <0.001.

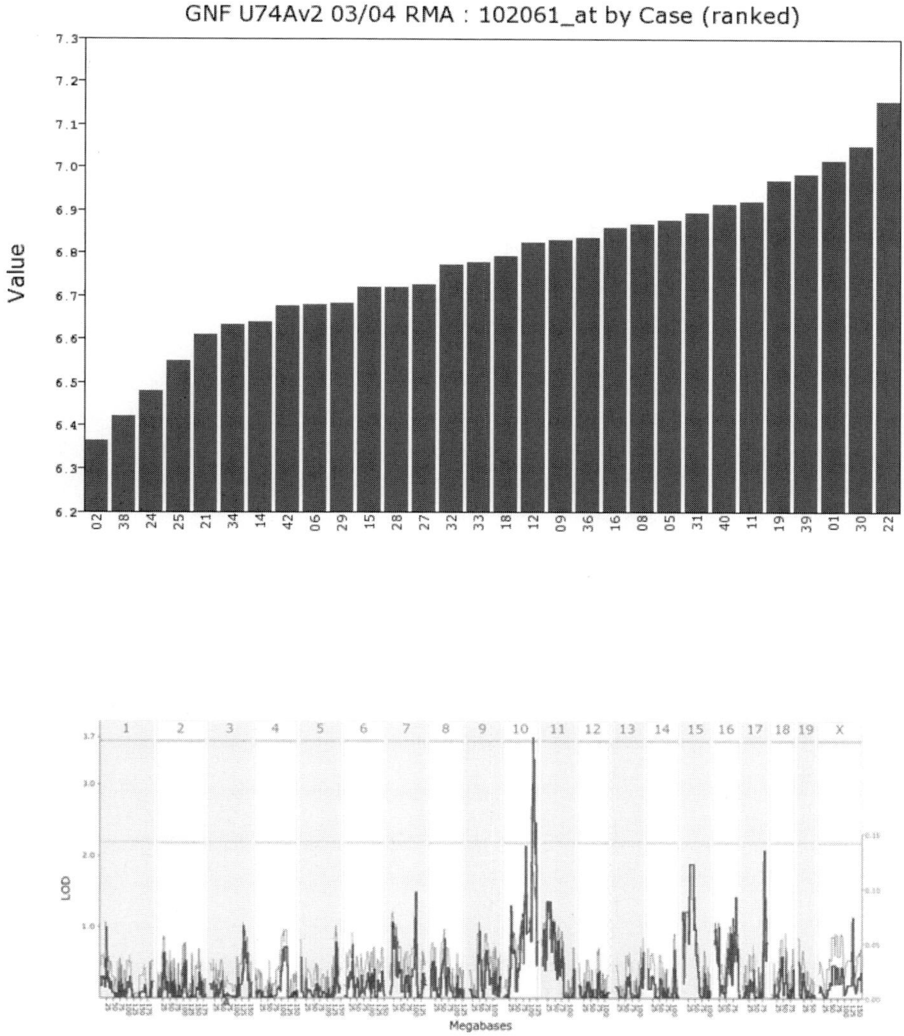

FIG. 1. Mapping genes regulating *Mlf1* expression. (Upper Panel) Variation in *Mlf1* transcript levels between HSC of 30 BXD strains. Expression data were generated using Affymetrix U74Av2 arrays and normalized using the Robust Multichip Average as described (Byskrykh et al 2005). Levels of *Mlf1* transcripts vary across a 10-fold range amongst the BXD strains tested. (Lower Panel) Genome-wide linkage analysis of *Mlf1* expression amongst BXD strains. Horizontal grey lines indicate significant (higher) and suggestive (lower) linkage thresholds, respectively. The location of *Mlf1* on chromosome 3 is arrowed. Clearly, the major locus affecting *Mlf1* expression in the BXD strains is located on chromosome 10. Further information displayed on the graph is as follows, but is not clearly visible in this black and white version. A positive additive coefficient (green line on the original) indicates that DBA/2J alleles increase trait values. In contrast, a negative additive coefficient (red line on the original; here dark grey) indicates that C57BL/6J alleles increase trait values. A copy of the original colour figure may be obtained by sending a request to G. M.

Multiple Mapping | Cluster Tree | Add Selection | Select All | Select None

Select Traits with r > -1.0 AND r < 1.0

☑ Display strain names in correlation plot
☑ Display fit line in correlation plot

	ProbeSetID	Symbol	Description	Chr	Position(Mb)	Correlation	#Strains	p Value
1 ☑	102061_at	Mlf1	myeloid leukemia factor 1	3	68.965	1.0000	30	0.00e+00
2 ☑	104157_at	Ubce7ip5	ubiquitin conjugating enzyme 7 interacting protein 5	2	130.756	0.6807	30	1.60e-05
3 ☑	99133_at	Slc3a2	antigen 4F2, heavy chain	19	8.598	-0.6514	30	5.32e-05
4 ☑	93765_at	Dcpp	demilune cell and parotid protein	17	23.511	0.6152	30	0.00019
5 ☑	92983_at	Ptprj	protein tyrosine phosphatase, receptor type, J	2	90.974	0.6111	30	0.00022
6 ☑	92375_at	Ascc1	activating signal cointegrator 1 complex subunit 1	10	61.924	-0.5985	30	0.00033
7 ☑	92639_at	Stk6	serine/threonine kinase 6	2	172.792	0.5962	30	0.00036
8 ☑	160583_at	Xlkd1	extra cellular link domain-containing 1	7	103.273	0.5790	30	0.00059
9 ☐	103405_at	2610019A05Rik	RIKEN cDNA 2610019A05 gene	11	105.827	-0.5786	30	0.00060
			immunoglobulin					

FIG. 2. A cluster of genes whose expression is correlated with that of *Mlf1*. WebQTL can return up to 500 genes whose expression is correlated with the gene of interest. In this example, transcripts of eight genes are shown which displayed positive or negative correlations with *Mlf1* expression, with $|r|$ values >0.5 and P values as low as 1×10^{-5}.

The next WebQTL application is very powerful: by integrating expression of each gene of this cluster with the underlying genotypes of the BXD strains, we can map loci throughout the genome which may influence expression of each of these genes. The output of this analysis is shown in Fig. 3. Each lane in this figure represents an individual transcript of the *Mlf1* cluster. These transcripts are clustered in a tree according to their correlation status. Information regarding the location of the gene encoding each of these transcripts is also provided, both by the text at the top of the figure and depicted by a triangle at the appropriate position within each lane. The Y axis represents each point in the genome, from chromosome 1 at the top through to the X chromosome. Here, the results are presented as a heat map, with individual effects of the parental B and D alleles on transcript levels indicated with different colours. Even with this 'helicopter view'

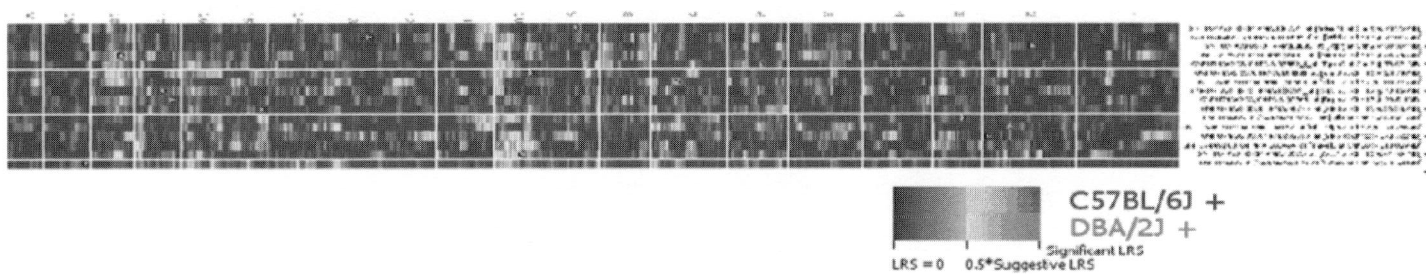

FIG. 3. Cluster tree analysis of 15 genes whose expression is correlated with *Mlf1* in HSC. See text for details. A copy of the original colour figure may be obtained by sending a request to G. M.

one can see (from the black/grey areas) that most of the genome has no effect on expression of any of these genes; this is of course to be expected. What is also immediately noticeable is that there are several brightly coloured areas, which indicate the position of loci at which strains with similar expression levels share more of the parental B or D alleles than could be expected by chance. That is, these are loci showing linkage to these phenotypes; these loci contain genes which influence the levels of the transcripts in the *Mlf1* cluster.

Focusing on specific regions allows us to formulate hypotheses that may be tested subsequently. For example, Fig. 4 shows the influence of loci on chromosome 3 on the levels of these transcripts. First, there is a weak association of a locus on chr 3 on the expression of *Mlf1* (arrowed). Perhaps this gene may be *Vannin3* (*Vnn3*), even though there seems to be no *cis*-effect on expression of *Vnn3* itself. More distally on chr 3 is another locus that is more strongly associated with changes in *Ptprj* expression. Levels of *Ptprj* mRNAs are strongly affected by a locus on chr 10; as *Ptprj* itself maps to this location, this is probably an effect of some polymorphism in its promoter or other *cis*-regulatory element. A number of other

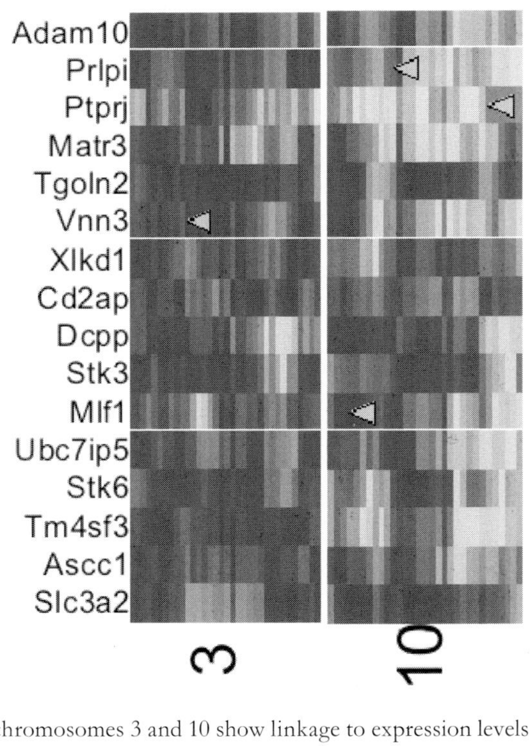

FIG. 4. Loci on chromosomes 3 and 10 show linkage to expression levels of transcripts from the *Mlf1* cluster. See text for details. A copy of the original colour figure may be obtained by sending a request to G. M.

genes in the cluster, including *Mlf1*, appear to be also regulated by this *Ptprj*-linked element. *Mlf1* maps to chr 3, but (consistent with the mapping results in Fig. 1) there are no *cis*-effects on its own expression. However, the *Mlf1* gene is weakly linked to differences in expression of another gene, *Tm4sf3*. Putting these results together, we start to see the emergence of a network of genes: a gene on distal chr 3 influences expression of *Ptprj*, which controls expression of a cluster of other genes, including *Vnn3* and *Mlf1*. In turn these genes appear to influence *Mlf1* and *Tm4sf3*, respectively. This hypothetical network contains clusters of transcripts as well as genes both upstream (e.g. the chr 3 locus) and downstream (*Tm4sf3*) of *Mlf1*. Thus, a simple WebQTL session generates hypotheses that can initiate investigations into genes previously unsuspected of having any functional involvement with the original reference gene.

These examples give an indication of only some of the functions available within GeneNetwork. Currently, systems genetics is a powerful technology for defining clusters of co-regulated genes. Its use is centred upon user-specified genes and can identify novel potential master regulatory genes for further investigation. We are working to increase the functionality and power of GeneNetwork and systems genetics in a number of areas. In particular, the mapping resolution can be increased by increasing the number of strains studied. By increasing the genetic diversity of the founders of an RI set, the potential for observing regulatory polymorphisms increases dramatically. In this context, the availability of 1000 RI strains from 'The Gene Mine' (a.k.a. the Collaborative Cross) (The Complex Trait Consortium 2004) will drive the development of systems genetics in further exciting areas.

Acknowledgements

This work was supported by grants from the National Health and Medical Research Council of Australia (#257601 and 305500), the University of Western Australia's Strategy Fund and the Diabetes Research Foundation [to GM]; and by NIAAA U01AA13499 and P20-DA 21131 [to RWW].

References

Bystrykh L, Weersing E, Dontje B et al 2005 Uncovering regulatory pathways that affect hematopoietic stem cell function using 'genetical genomics'. Nat Genet 37:225–232

Chesler EJ, Lu L, Wang J, Williams RW, Manly KF 2004 WebQTL: rapid exploratory analysis of gene expression and genetic networks for brain and behavior. Nat Neurosci 7: 485–486

Chesler EJ, Lu L, Shou S et al 2005 Complex trait analysis of gene expression reveals polygenic and pleiotropic networks that modulate nervous system function. Nat Genet 37: 233–242

Hubner N, Wallace CA, Zimdahl H et al 2005 Integrated transcriptional profiling and linkage analysis for identification of genes underlying disease. Nat Genet 37:243–253

Li H, Lu L, Manly KF et al 2005 Inferring gene transcriptional modulatory relations: a genetical genomics approach. Hum Mol Genet 14:1119–1125

Peirce JL, Lu L, Gu J, Silver LM, Williams RW 2004 A new set of BXD recombinant inbred lines from advanced intercross populations in mice. BMC Genet 5:7

Taylor BA 1989 Recombinant inbred strains. In: Lyon ML, Searle AG (eds) Genetic variants and strains of the laboratory mouse 2nd Ed. Oxford University Press, p 773–796

The Complex Trait Consortium 2004 The collaborative cross. Nat Genet 36:1133–1137

Wang J, Williams RW, Manly KF 2003 WebQTL: web-based complex trait analysis. Neuroinformatics 1:299–308

Williams RB, Cotsapas CJ, Cowley MJ, Chan E, Nott DJ, Little PF 2006 Normalization procedures and detection of linkage signal in genetical-genomics experiments. Nat Genet 38:855–856

Yoneda-Kato N, Look AT, Kirstein MN et al 1996 The t(3;5)(q25.1;q34) of myelodysplastic syndrome and acute myeloid leukemia produces a novel fusion gene, NPM-MLF1. Oncogene 12:265–275

DISCUSSION

Wakeland: I get the impression that what you are describing is being looked at in the context of these genes all being expressed in a single cell, with a master gene regulating the expression of other genes in the cell. But really what you are looking at is a tissue that has a variety of cell subsets within it. As you go from strain to strain and look at various points on the chromosome, this may either be affecting the expression of the gene, or the presence of a particular cell subset in various ratios. One of the issues I've had with doing spleen or thymus, for example, is that there is a very complex mixture of cells. The actual proportions of these cells vary among strains. Variations in gene expression may therefore reflect either changes in transcription or changes in the populations of cells present in the starting sample.

Morahan: That's true, but it is also a strength of this system. You are not limited to looking at interactions just within the same cell but potentially a number of different genetic mechanisms. The important point is that this is a way of forming hypotheses about genetic relationships that you can then test experimentally.

Wakeland: What would appear to be a master regulator could be a cell that impacts the level of expression of a particular cell type or lineage. The outcome would be a whole series of genes that could go up or down, not based on this being a regulator at all, but simply impacting in some potentially non-important way the proportion of cells present in tissue.

Hume: I think what Grant Morahan says about this being a benefit can also work. We did a somewhat similar analysis looking at the gene expression profiles of several hundred breast cancers. The macrophage content of the breast cancers varies substantially from 10–70%. You can do a correlation analysis with any gene you are interested in, say the CSF-1 receptor, and ask what genes are correlated with the expression of that gene. You can tell that there is a cluster that is every known macrophage-specific gene. So the dominant variable is in fact the number of macrophages. But the other genes in that cluster enable you to phenotype the macrophages of the breast cancer. In some measure, if the stem cell content varies in the different preparations, or the macrophage or neutrophil content, you get a *de facto* phenotyping of that cell type just because of the variation in abundance. I don't know that you can infer causal relationships, but this association leads to hypotheses. It is powerful.

Foote: We have done quite a lot of transcriptional profiling in congenics. One of the striking findings is that many of the genes are *cis*-regulated. There is a lot of genetic variation in the *cis*-regulatory elements, with a lot of differential expression between strains based on congenic intervals.

Morahan: I have seen many examples using this methodology when the interactions have not been in *cis*.

Foote: I have seen a few presentations where a lot of the interactions have been *cis* interactions. We see an enormous amount of differential expression between different congenic intervals.

Morahan: It is logical that many genes will be regulated differentially by polymorphisms in the gene itself, or associated with it. Of course there will be many examples of such *cis*-regulators as they are the obvious ones to find, and can be seen using conventional technologies and resources like congenic mice. Therefore, *cis*-factors will dominate from the conventional studies that you describe. However, this systems genetics approach is able to define both *cis* and *trans* regulators. From the studies you mention, it would have to be said that it is more striking to find *trans*-acting regulators and we see many examples of these, even in the absence of *cis* effects. These could of course be due to indirect (e.g. extracellular) effects such as those mentioned by David Hume.

Goodnow: The *cis* differences would be better picked up if you could measure single nucleotide polymorphisms (SNPs) within transcripts. Then you would only need to measure the relative abundance of transcript SNPs in a single F1 hybrid. If this is what you were after, it would be a much more powerful way to get at the *cis*-acting differences than expression. But if you want the *trans*-acting differences—for example, if you suspect there is an amino acid polymorphism in a transcription factor or signalling pathway—the issue is going to be teasing this apart from the one that Ward Wakeland mentioned. We fell foul of this when we were microarray profiling B cells from two different transgenic strains that were closely matched, but where the level of purity of the B cells varied slightly. They were 95% pure from one strain and 85% pure from another. The strategy we used was to deplete most of the other contaminating cell types except for red cells and neutrophils. We had a bunch of genes that came up as up-regulated in one B cell type. Among them were red cell-specific highly abundant genes or neutrophil genes. The only way to sort out these were not up-regulated in the cell type of interest was to FACS (fluorescence-activated cell sorting) highly purified cells.

Hume: The *cis*-acting ones should be fairly obvious because the linkage of the expression will locate to the locus itself. It won't correlate all that well: if it is purely *cis*-acting and there are no other consequences there will be no other linkage, if you treat each gene expression essentially as a quantitative trait.

Morahan: They do stand out: *cis*-regulators are the obvious type to find, particularly using existing resources. They can certainly be found with this approach, but we commonly see individual genes whose gene expression is not regulated by variation in their own alleles, but by variation in *trans*-acting genes.

Fortin: A study was conducted by Lee et al (2006) in Thomas Husdon's laboratory at McGill University using our set of recombinant congenic strains. Gene expression profiling was performed with lung RNA from the strains across our panel. Using a multiple-regression model measuring the association between gene expression level, donor strain of origin (DSO) and predominant strain background, they identified over 1500 genes ($P < 0.05$) whose expression profiles differed according to the DSO. *Vanin3* was a clear example of such differentially expressed genes.

Wakeland: We work with a lot of investigators doing various types of experiments looking at transcriptional data. I agree with Simon Foote: we see an awful lot of genes in congenic strains that are up-regulated from the interval. We can certainly show that they correlated by quantitative RT-PCR, so it is a legitimate, significant polymorphism in transcription that we see, due to variations in transcription that are *cis*. Beyond that, as Chris said, the biggest issue is the relative changes in the cellular content of samples that are trying to be viewed as being identical. The outcome can be that we see big chunks of genes going up and down in an apparently exciting and then confusing fashion. The more we know about the input sample, the more feasible it is to try to understand what is going on. If you know neutrophils are a potential contaminant, then you can begin to sort the data on this basis.

Hume: That is why we have always worked on pure macrophages grown out of the animal at the greatest purity we can get.

Gros: One potential issue is the effect of other mutations in the genome, not necessarily in transcription factors, but in other proteins or pathways that will have repercussions on cell populations. This will show up on the array analysis as behaving as a *trans*-acting factor. The PKLR mutation, for example, has a dramatic effect on spleen. Half of the genes in the interval appear to be regulated in this way. It is just because there is a change in cell populations. The other pitfall is this. If you were to throw the C5 mutation of A/J into a BXD cross, how would this behave in terms of controlling downstream targets? We know that the C5 mutation has a pleiotropic effect on a large number of cytokines and cell responses of different pathways. This would map as a master regulator. How do we deal with such possible pleiotropic effects if we are not aware of them in the first place If you map such an effect back to a transcription factor you almost need to set up an experiment where you fuse that transcription factor to, say, a heart transcriptional activator and do a correlation. You need some sort of validation.

Morahan: Of course, we need to validate it, but it is a way we can identify novel genes rapidly that may be in the same genetic network as the gene you are interested in. Then you need to validate the role of that gene and to identify its function in that network. The point is this is a powerful methodology that can provide data in half an hour that allows you to form hypotheses that you can then spend years investigating.

Reference

Lee PD, Ge B, Greenwood CM et al 2006 Mapping cis-acting regulatory variation in recombinant congenic strains. Physiol Genomics 25:294–302

Regulation of the immune system in metazoan parasite infections

Rick Maizels

Institute of Immunology and Infection Research, University of Edinburgh, West Mains Road, Edinburgh EH9 3JT, UK

Abstract. Eukaryotic, multicellular parasites such as the helminth worms have a major impact on the mammalian immune system in two contexts. First, they have evolved sophisticated strategies for long-term immune evasion including recruiting natural suppressive mechanisms such as the regulatory T cell (T_{regs}). T_{regs} play a role not only in repressing immunity to parasites, but also in dampening bystander responses such as those to allergens. To achieve these effects, they produce a range of immunomodulators some of which are evolutionary homologues of immune system cytokines, while others are novel proteins capable of interfering with immune cell signalling and differentiation. The second context in which metazoa may have influenced their host is at the level of genetic polymorphism in immune response genes. Alleles at loci originally associated with predisposition to asthma have more recently been found to confer heightened resistance to helminth parasites. This may suggest a mechanistic link between more vigorous type 2 responses in both allergy and infection. On a broader perspective, one may speculate that alleles advantageous in the historical environment of prevalent infection, now display a deleterious phenotype in our more 'hygienic' societies.

2007 Decoding the genomic control of immune reactions. Wiley, Chichester (Novartis Foundation Symposium 281) p 192–207

Helminths are multicellular worm parasites which today infect well over one quarter of the world's population (de Silva et al 2003). Helminth parasitism would have been a constant feature of the pathogen environment within which the modern immune system evolved. The continuing high prevalence of helminth infections reflects their ability to survive in the human body for many years, apparently unscathed by the host immune system. This feat is made the more remarkable by the fact that in filariasis and schistosomiasis, the parasites home to lymphatic and vascular vessels respectively, remaining in an extracellular milieu accessible to the full armoury of the immune system. A further striking feature of helminth-mediated diseases is the spectral nature of both the host immune response and the pathological outcomes. For example, in lymphatic filariasis, the outcome of exposure can vary from no evident infection (implying resistance) to

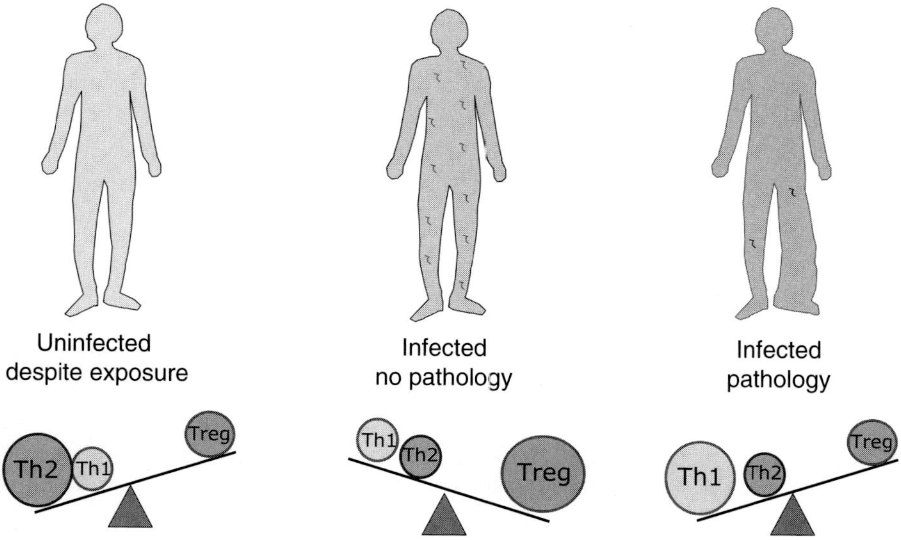

FIG. 1. The spectrum of helminth-induced disease may reflect different degrees of regulatory T cell activity. Those with effective immunity (left) mobilize an appropriate (Th2-dominated) immune response. Carriers may accumulate high level infections (centre) but are protected from immunopathology by regulatory T cells. Cases which progress to pathology (right, depicting elephantiasis due to filarial nematodes) can clear most of the parasites but incur pathological damage, attributable to inappropriate immune responsiveness.

chronic oedema and elephantiasis, although the most common status is that of an asymptomatic microfilaraemic carrier (Fig. 1). Intriguingly, parasite numbers (measured by microfilariae in the blood) are highest in the asymptomatic carriers, suggesting that some form of imunological tolerance to the parasite has been established (Maizels & Yazdanbakhsh 2003). In contrast, more immunologically reactive individuals display stronger responses, fewer parasites and overt pathology in the form of lymphatic damage, lymph stasis and secondary infections.

The immune response to helminth infection is generally strongly polarized towards the Th2 phenotype, with a low level of Th1-type components (Pearce et al 2004). While the Th2 response is widely associated with protection from helminth infection (Grencis 2001), there is a paradoxical failure of immunity in many individuals with a dominant Th2 phenotype. However, when analysed in more detail it is evident that the 'tolerant' individual mounts a muted Th2 response in which inducer/regulatory cytokines interleukin (IL) 4 and IL10 are produced, but the key product IL5 (which mediates eosinophil and granulocyte inflammatory responses) is down-modulated. Moreover, in carriers with high microfilarial levels there is a skewing of antibody isotypes away from IgE and toward IgG4, which is explained by the differential effects of IL10 in controlling these two Th2-associated isotypes. Thus infection selectively reduces both effector cytokine (IL5) and inflammatory antibody (IgE) levels, producing a profile described in the allergic context as the 'modified Th2' response (Platts-Mills et al 2001).

Because the spectrum of filarial disease does not readily conform to a simple Th1/Th2 dichotomy, and because there is selective down-modulation within the Th2 compartment itself, it has recently been suggested that the determining factor in immunity is not the presence or absence of a Th1 component, but the activity of immunosuppressive regulatory T cells (Hoerauf & Brattig 2002, Maizels & Yazdanbakhsh 2003). This proposition can account for the paradox of Th2 dominance without Th2 effector function, and as discussed below is in accord with a growing body of experimental and epidemiological evidence. If confirmed, we will have to acknowledge that helminths have managed to exploit the ultimate Achilles heel of the immune system, by recruiting into their service, the regulatory T cells (T_{regs}) whose normal responsibility in the immune system is to prevent autoimmunity (Sakaguchi 2004).

Emergence of regulatory T cells

The past decade has seen the re-establishment of the concept of suppressive T cell populations in a new guise, T_{regs} (Sakaguchi 2004, Shevach et al 2001). These cells were first defined as natural inhibitors of autoimmunity, emerging from the thymus with self-specificity to maintain immune homeostasis. These T_{regs} constitutively express CD25 (IL2Rα) on their surface, and the transcription factor Foxp3 at the intracellular level (Hori et al 2003). In addition to naturally developing T_{regs} two 'inducible' types are recognized: Th3, activated in mucosal tissues, releasing soluble suppressive cytokines (IL10 and transforming growth factor [TGF]β); and Tr1 which is principally associated with IL10 production. Th3 and Tr1 cells do not necessarily express CD25 or Foxp3, but at the present time it is not known whether these phenotypes represent exclusive or interconvertible populations. Most regulatory cells express high levels of CTLA4, which interferes with the CD28-dependent co-stimulatory pathway on normal T cells, and may also act on dendritic cells (DCs) to induce a 'regulatory DC' phenotype (Mellor et al 2004). Regulatory T cells are also associated with high expression of the surface receptor GITR (Shimizu et al 2002).

A central question for our understanding of T_{reg} origins is the role of non-self antigens, not present in the thymus, in the development of T_{reg} activity (Bluestone & Abbas 2003). The finding that T_{regs} prevent reactivity to food antigens and commensal bacteria (Powrie 1995), establishes that T_{reg}-like cells can adaptively develop to regulate exogenous specificities. Further, T_{regs} can play an important role in the response to viral (Suvas et al 2003), bacterial (Kullberg et al 2002, McGuirk et al

2002), protozoal (Belkaid et al 2002, Hisaeda et al 2004) and most recently helminth (Taylor et al 2005) infections. However, it is still not clearly established whether pathogen-reactive T_{regs} are in fact 'natural' T_{regs} stimulated by parasite epitopes (as reported in *Leishmania major* infections; Suffia et al 2006), or can be generated from naive Th0 cells in the periphery.

Immune down-regulation in Helminth infection

Several key features of the down-modulated immune response to helminth infection are consistently observed in studies with a number of filarial and schistosome species (Maizels & Yazdanbakhsh 2003). There is, firstly, a loss of antigen-specific proliferative responsiveness by peripheral blood T cells *in vitro*, which can be restored by curative drug treatment or, in some cases, by adding neutralizing antibodies to IL10 or TGFβ to antigen-stimulated cultures (King et al 1993). Thus, the presence of parasites correlates with depressed proliferation, but patients do not lack the ability to recognize parasite antigens under favourable conditions. The latter point is reinforced by the fact that while antigen-stimulated T cells fail to mount a proliferative response, they can respond by production of IL4 and IL10. Critically, both IFNγ (a Th1 inflammatory cytokine) and IL5 (which can be considered a Th2 inflammatory cytokine) are suppressed during infection (Grogan et al 1998, Sartono et al 1997). As with antigen-specific proliferation, IL5 responsiveness can be rescued by drug treatment of aysmptomatic carriers (van den Biggelaar et al 2002). Thus, antigen-reactive T cells are present in infection but their activity is circumscribed to the counter-inflammatory cytokines; moreover, the suppression of both IFNγ and IL5 does not indicate that the alterations in helminth infection represents a Th1–Th2 antagonism, since IL4 and IL5 are differentially regulated.

If immune modulation in infection is not based upon the Th1/Th2 balance, is there a significant component of T_{reg} activity? Evidence for elevated constitutive IL10 expression in human filariasis (Mahanty et al 1996), together with the rescue of *in vitro* proliferation with anti-IL10 and TGFβ (King et al 1993), gave early indications that this was the case. More recently, T cell clones expressing the same two regulatory cytokines, and high levels of surface CTLA4 have been isolated from onchocerciasis patients (Doetze et al 2000, Satoguina et al 2002). Elevated CTLA4 expression was also noted in lymphatic filariasis patients, particularly marked on the $CD4^+CD25^+$ population of microfilaraemic carriers (Steel & Nutman 2003). Significantly, blocking CTLA4 function *in vitro* resulted in increased IL5 responses from peripheral T cells, again pointing to a regulatory link suppressing IL5 in human filariasis. The association between active filarial infection and expanded T_{reg} activity has been most recently demonstrated at the level of expression not only of the Foxp3 transcription factor, but also a suite of T_{reg}-associated molecules such as indoleamine-2,3-dioxygenase (IDO), ICOS and PD1 (Babu et al 2006).

T_{regs} maintain susceptibility to filarial nematode infection

The principal filarial parasites of man, such as *Brugia malayi*, survive for only short periods in laboratory mice, and experimental studies are therefore performed on model system species. One such model is *Litomosoides sigmodontis*, phylogenetically close to *Brugia*, and a natural parasite of rodents. Infection is transmitted by arthropod mite vectors; once introduced into the mammalian body, larval parasites migrate through the lymphatics to the thoracic cavity where they mature, mate and produce live microfilarial offspring. Mouse strains vary in their degree of resistance to infection, with C57BL/6 being able to eliminate parasites before the production of microfilariae, while BALB/c are fully susceptible and display patent infection. Resistance in the C57BL/6 strain is achieved through a strong Th2 response, as IL4-deficient mice are fully susceptible (Le Goff et al 2002). However, susceptible BALB/c animals also mount a polarized Th2 response, giving rise to the hypothesis that their failure to clear parasites is due to the activity of T_{reg} cells.

To support this hypothesis, we found rapid expansion of $Foxp3^+CD25^+$ T cell numbers in the thoracic cavity and its draining lymph nodes soon after infection, as well as regulatory cytokine (IL10 and TGFβ) production in infected mice. Over the course of infection, there was a more balanced growth in both $Foxp3^+CD25^+$ T_{reg} cells, and $Foxp3^-CD25^-$ T cells which would normally be considered effector cells. Notably, the $Foxp3^-CD25^-$ population showed increasing levels of GITR and CTLA4 surface expression over time, reaching a maximum at 60 days post-infection, by which time this the 'effector' population was hyporesponsive or anergised, having lost its ability to mount parasite antigen-specific IL5 responses (Taylor et al 2005).

Antibody-mediated depletion of CD25-expressing cells *in vivo* was then used to analyse the role of T_{regs} in infection. Anti-CD25 antibody alone, given after 4 weeks of infection, had no effect on the ability of parasites to survive in BALB/c mice despite removing the $CD25^+$ T cell population *in vivo*. However, when anti-CD25 was combined with either an agonistic antibody to GITR, or a blocking antibody to CTLA4, mice were able to clear the majority of parasites. The combined antibody treatment also restored T cell IL5 responses to parasite antigens. Because the anti-CD25 antibody effectively cleared the T_{reg} population, and as the hyporesponsive $GITR^+CTLA4^+$ population had regained cytokine reactivity, we proposed that the second antibody (to GITR or CTLA4) had its effect not on T_{reg} but on the anergized population, restoring them to responsiveness. Thus, nullifying T_{reg} activity can 'cure' chronic helminth parasite infection, but only if there is restoration of immune responsiveness through manipulating the co-stimulatory network of the effector population (Taylor et al 2005).

The requirement for combined treatment with anti-CD25 and antibodies to markers involved in co-stimulation is noteworthy, and suggests that T_{regs} maintain a state of hyporesponsiveness among the effector T cell population. Effector Th2 cells may therefore be in a 'conditioned' or 'modified' state, exactly as in human helminth infections where T cells are hyporesponsive at the proliferative level, yet mount intact IL4 and IL10 responses, and possibly akin to the 'modified Th2' condition observed in desensitized allergy patients (Platts-Mills et al 2001). Most recently, the hyporesponsive human T cell population has been shown to have up-regulated a set of anergy-inducing gene products (including Cbl-b, c-Cbl, Itch and Nedd4 (Babu et al 2006), offering a new molecular window onto the immunological pathways to suppression in human helminth disease.

Helminth infections and modulation of allergies

Arguably the two most prominent settings for the Th2 response are helminth infections and allergic diseases, each characterized by stimulation of IgE and eosinophilia. Helminthiases might, therefore, be expected to exacerbate allergies, but the weight of epidemiological and experimental evidence points to an inverse relationship between infection and overt allergy. Allergic diseases are much less prevalent in countries with high helminth endemicity, and even within endemic areas atopy is less frequent in individuals with chronic worm infections; moreover, allergic responses are heightened in infected individuals given curative anthelminthic treatment (reviewed by Maizels 2005). The involvement of the immune regulatory network in this interaction was suggested by a study in Schistosome-infected Gabonese schoolchildren, in which protection from skin atopy correlated with parasite-specific production of IL10 (van den Biggelaar et al 2000). Other investigators showed that both allergen- and parasite-specific IL10 responses were higher in infected subjects, but abated following drug-induced parasite clearance (Araujo et al 2004). Thus, the presence of a helminth infection promotes regulatory cytokine production in response to a structurally unrelated allergen. However, human studies cannot resolve whether the cross-inhibition of allergy is due to a regulatory T cell effect or a different type of interaction such as antigenic competition or alteration of the immune environment by migrating parasites.

To address these questions in an experimental setting, we infected mice with a gastrointestinal parasite (*Heligmosomoides polygyrus*) which does not traverse the pulmonary tissues, and tested levels of airway inflammation using Th2-biasing immunization protocols for stimulating allergy. We found a sharp reduction in airway cellular infiltration and inflammation (such as eosinophilia and goblet cell proliferation) in mice carrying a chronic *H. polygyrus* infection. These results were reproduced with two allergens (Ova and Der p1) in two mouse strains (BALB/c and C57BL/6), eliminating the possibility that fortuitous cross-reactivity between an allergen and a parasite was responsible for the effect. The reduction in allergy was not accompanied by any switch from Th2 response to the Th1 mode, as IL4 levels remained high and IFNγ low in both uninfected and infected mice. However, IL5 and IL13, as well as eotaxin, were depressed in infected animals, suggesting perhaps that while Th2 'inducer' cytokine IL4 was unaffected, the 'effector' cytokines IL5 and 13 were selectively suppressed in infection (Wilson et al 2005).

It was likewise shown that infection does not inhibit the sensitization phase of the response, as allergen-specific antibody levels were unaffected by the presence of the infection. To demonstrate this conclusion formally, we primed mice with allergen prior to infection, and found that infected pre-sensitized animals had suppressed allergic airway inflammation. Thus, the presence of infection induces an immunological component capable of down-regulating a pre-existing memory population.

We hypothesized that this component would be the regulatory T cell. To support this contention, we demonstrated activation of the $CD25^{+}Foxp3^{+}$ population within the mesenteric lymph nodes of *H. polygyrus* infected mice (Wilson et al 2005), and the enhanced suppressive capacity of $CD25^{+}$ MLNC T cells in blocking mitogen-induced *in vitro* proliferation of $CD25^{-}$ target cells (Finney, unpublished). We also showed that the down-modulation of allergy in infection was abolished in mice treated with depleting anti-CD25 antibody. As it had been shown that infection can suppress a previously primed allergic response, we then conducted cell transfer studies with MLNC injected into allergen-sensitized but parasite-naïve recipients.

MLNC from infected, but not uninfected, mice were found to suppress airway allergy in primed animals tested 7 days after transfer. Fractionation of the MLNC prior to transfer showed maximal suppression was associated with $CD4^{+}CD25^{+}$ T cells, although other populations can also confer an effect, particularly in the C57BL/6 strain (Wilson et al 2005). Interestingly, the transferred T cells did not require re-stimulation with antigen in their adoptive host. This fact could reflect a high degree of activation within the infected donor environment, or else suggest that the T_{regs} are specific for a self antigen present in both donor and recipient animals.

The role of IL10 in regulation of Th2 responses

IL10 was first characterized as a Th2 cytokine capable of inhibiting Th1 responsiveness, but in more recent years its role as a suppressive product of T_{reg} cells has become more prominent. In the context of helminth infections, it is important to note that IL10 acts from an early stage in promoting Th2 dominance, and where the Th2 response is protective, IL10 may be an essential component of the protective immune response (Helmby & Grencis 2003). This double-edged nature of IL10 is also observed in the control of pathology; for example, in murine schistosomiasis, IL10 protects the host from egg-induced pathology mediated in different ways by both Th1 and Th2 cells (Hoffmann et al 2000). Thus, within the Th2 orbit, IL10 is intimately involved in both induction and repression of activity.

We therefore tested the role of IL10 in the model systems of immunity and allergy described above. In *L. sigmodontis*-infected mice, antibody neutralization of IL10 signalling *in vivo* had no effect on the outcome of infection, other than to raise the overall level of Th1 responsiveness observed to the parasite. In *H. polygyrus* infections, there is a marked expansion in IL10-producing T cells, but the majority of such cells are $CD25^-$. Indeed, it would appear that in a typical Th2 response, most IL10 is derived from Th2 cells rather than T_{regs}. Significantly, suppression of allergic inflammation is unaltered in animals in which IL10 signalling is blocked, indicating that mediators other than IL10 are responsible for the infection–allergy interaction. Moreover, suppression of allergy can be transferred with cells from infected IL10-deficient animals. It therefore seems likely that other suppressive mediators (e.g. TGFβ) are pre-eminent in *H. polygyrus* infection, and that suppression of Th2 effector responses in anti-helminth immunity or airway allergy may be less susceptible to IL10-mediated suppression than Th1 inflammatory responses.

Molecular mediators of parasite immune regulation

A major challenge for the future will be to identify functional mediators from parasites which act to induce down-regulatory activity within the host immune system. Although beyond the scope of this article to describe at any length, a number of important protein molecules thought to be responsible for immune evasion have now been identified (Maizels et al 2001, Maizels & Yazdanbakhsh 2003), as well as key non-protein constituents such as phosphorylcholine (Harnett et al 2005). Some of these, such as serine protease inhibitors, target specific host enzymes and the co-evolution of parasite and host partners has been likened to a molecular arms race (Zang & Maizels 2001). In filarial nematodes, two gene families related to mammalian cytokines have been found: these are homologues of macrophage migration inhibitory factor (MIF; Zang et al 2002) and of TGFβ (Gomez-Escobar et al 2000). In the latter case, secretion of a parasite TGFβ homologue able to bind mammalian TGFβ receptors may be a significant factor in the induction of T_{reg} populations during infection.

In addition to parasite genes with clear sequence homology to known gene families, many prominent helminth products are completely novel structures. One example is found in filarial infective larvae, in which up to 5% of cDNA corresponds to a small (~125 aa) protein termed abundant larval transcripts (ALTs; Gomez-Escobar et al 2005). To test the hypothesis that ALT proteins interfere with host immunity, we transfected these genes into the protozoal parasite *Leishmania mexicana*, and demonstrated both *in vitro* and *in vivo* that ALT expression results in faster growth within macrophages as well as higher survival *in vivo*. At the molecular level this property corresponded with up-regulation of SOCS1 within macrophages (Gomez-Escobar et al 2005), which by interfering with IFNγ signalling is likely to impede the ability of the host cell to attack parasites.

Parasites and the evolution of the immune system

The coordinate evolution of the immune system with the universe of pathogens is likely to have extended beyond a molecular arms race. In particular, if parasites such as helminths exert a modest but significant immunodepressive effect, our immune system will have compensated in order to attain an optimal level of immunological activity. Thus, in the absence of parasites it may be that the immune system will tend to over-react, possibly sparking immunopathological outcomes such as allergy and autoimmunity (Fig. 2). If so, then loci controlling susceptibility to parasite infection may also influence propensity to develop immunopathological diseases in modern environments.

It is also likely that, confronted with an ever-changing pathogen milieu (in which for example, low intensity infections demand a different optimal response to high intensity), the immune system has retained a large number of balanced polymorphisms in immune response genes. To compensate for pathogen depression of immunity, such polymorphisms may take the form of quantitative rheostats on the level of gene expression, altering promoter efficiency or message longevity, rather than coding alterations in structural genes.

Evidence for the first proposition, that parasites may have driven evolution of immune system polymorphisms, is now emerging. In the first example, the -1055T allele of IL13 showed a significant association with allergic asthma (van der Pouw Kraan et al 1999), presumably because it regulates the quantitative expression of IL13 production. More recently, the same allele was shown to protect against Schistosomiasis: homozygous (TT) carriers showed lower worm burdens, as estimated by egg counts less than half those of CC homozygotes, particularly through the teenage years when transmission is highest (Kouriba et al 2005). Similarly a STAT6 3'UTR polymorphism, the same allele predisposing to asthma in Westerners, confers resistance to *Ascaris* worms in China (Peisong et al 2004). Such variants are presumed to effect subtle changes in the level of responsiveness, and it will be fascinating to see in coming years whether many of the other quantitative trait alleles of the immune system associated with asthma (Symula et al 1999) or

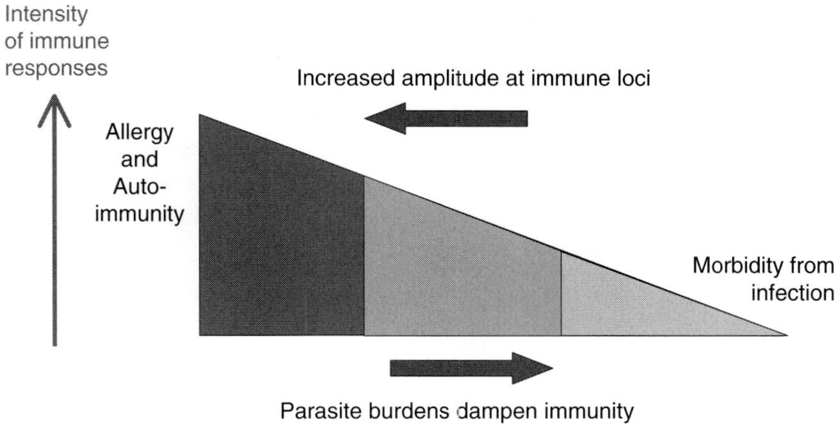

FIG. 2. Co-evolution of the immune response in a parasite-rich environment may have resulted in compensation for moderate immune suppression by parasites. Variants at immune loci may have evolved to up-regulate inflammatory gene expression (or down-shift regulatory gene expression), to optimize immune function in the presence of parasites and avoid deleterious effects of infection. An ever-changing infection landscape (parasite species, antigen variants, transmission intensities) is likely to have maintained polymorphisms at these loci. In modern environments with a lower presence of pathogens, certain genotypes could be predisposed to 'overshoot' in immune intensity as found in allergies and autoimmune diseases.

autoimmunity may have their origins in calibrating the response to pathogens. More speculatively, such loci may influence the level of regulatory T cell function—for example, among the loci controlling type 1 diabetes (T1D) in both human and mouse are CD25 (Vella et al 2005) and CTLA4 (Ueda et al 2003).

The fact that immune system loci such as these remain polymorphic may be a significant reflection of the diversity of pathogen assault that must be contained. A striking demonstration of this emerges from a study of a feral, closed sheep population which has bred without human intervention for many decades on the isolated island of St Kilda. Within this population, gastrointestinal nematode infection is a major cause of mortality, and resistance is associated with one allele linked to the IFNγ locus (Coltman et al 2000). Drawing a parallel to the rodent system, this allele is suggested to be a low expressing variant. However, the resistant allele has not achieved fixation, indicating that it is subject to balancing selection against a different pathogen, postulated to be a respiratory microbe which may demand a higher level of IFNγ response for protection.

While the specific effects and evolutionary causes for these immune system polymorphisms remain largely speculative at this time, the recent convergence of studies on helminth infections, allergies and human variation promise to reveal many fascinating insights in the near future. In particular, the ability of helminths to exploit the immune regulatory network provides a new perspective to probe and understand the normal pathways of immunological controls, as well as the potential to intervene rationally and cure infections by modulating the immune response to these sophisticated pathogens.

References

Araujo MIAS, Hoppe B, Medeiros M Jr et al 2004 Impaired T helper 2 response to aeroallergen in helminth-infected patients with asthma. J Infect Dis 190:1797–1803

Babu S, Blauvelt CP, Kumaraswami V, Nutman TB 2006 Regulatory networks induced by live parasites impair both Th1 and Th2 pathways in patent lymphatic filariasis: implications for parasite persistence. J Immunol 176:3248–3256

Belkaid Y, Piccirillo CA, Mendez S, Shevach EM, Sacks DL 2002 $CD4^+CD25^+$ regulatory T cells control *Leishmania major* persistence and immunity. Nature 420:502–507

Bluestone JA, Abbas AK 2003 Natural versus adaptive regulatory T cells. Nat Rev Immunol 3:253–257

Coltman DW, Wilson K, Pilkington JG, Stear MJ, Pemberton JM 2000 A microsatellite polymorphism in the gamma interferon gene is associated with resistance to gastrointestinal nematodes in a naturally-parasitised population of Soay sheep. Parasitology 122:571–582

de Silva NR, Brooker S, Hotez PJ, Montresor A, Engels D, Savioli L 2003 Soil-transmitted helminth infections: updating the global picture. Trends Parasitol 19:547–551

Doetze A, Satoguina J, Burchard G et al 2000 Antigen-specific cellular hyporesponsiveness in a chronic human helminth infection is mediated by T_h3/T_r1-type cytokines IL-10 and transforming growth factor-β but not by a T_h1 to T_h2 shift. Int Immunol 12:623–630

Gomez-Escobar N, Gregory WF, Maizels RM 2000 Identification of *Bm-tgh-2*, a filarial nematode homolog of *C.elegans daf-7* and human TGF-β, expressed in microfilarial and adult stages of *Brugia malayi*. Infect Immun 68:6402–6410

Gomez-Escobar N, Bennett C, Prieto-Lafuente L, Aebischer T, Blackburn CC, Maizels RM 2005 Heterologous expression of the filarial nematode *alt* gene products reveals their potential to inhibit immune function. BMC Biol 3:8

Grencis RK 2001 Cytokine regulation of resistance and susceptibility to intestinal nematode infection—from host to parasite. Vet Parasitol 100:45–50

Grogan JL, Kremsner PG, Deelder AM, Yazdanbakhsh M 1998 Antigen-specific proliferation and interferon-gamma and interleukin-5 production are down-regulated during *Schistosoma haematobium* infection. J Infect Dis 177:1433–1437

Harnett W, Goodridge HS, Harnett MM 2005 Subversion of immune cell signal transduction pathways by the secreted filarial nematode product, ES-62. Parasitology 130 Suppl: S63–68

Helmby H, Grencis RK 2003 Contrasting roles for IL-10 in protective immunity to different life cycle stages of intestinal nematode parasites. Eur J Immunol 33:2382–2390

Hisaeda H, Maekawa Y, Iwakawa D et al 2004 Escape of malaria parasites from host immunity requires $CD4^+CD25^+$ regulatory T cells. Nat Med 10:29–30

Hoerauf A, Brattig N 2002 Resistance and susceptibility in human onchocerciasis–beyond Th1 vs. Th2. Trends Parasitol 18:25–31

Hoffmann KF, Cheever AW, Wynn TA 2000 IL-10 and the dangers of immune polarization: excessive type 1 and type 2 cytokine responses induce distinct forms of lethal immunopathology in murine schistosomiasis. J Immunol 164:6406–6416

Hori S, Nomura T, Sakaguchi S 2003 Control of regulatory T cell development by the transcription factor *FOXP3*. Science 299:1057–1061

King CL, Mahanty S, Kumaraswami V et al 1993 Cytokine control of parasite-specific anergy in human lymphatic filariasis. Preferential induction of a regulatory T helper type 2 lymphocyte subset. J Clin Invest 92:1667–1673

Kouriba B, Chevillard C, Bream JH et al 2005 Analysis of the 5q31-q33 locus shows an association between IL13-1055C/T IL-13-591A/G polymorphisms and *Schistosoma haematobium* infections. J Immunol 174:6274–6281

Kullberg MC, Jankovic D, Gorelick PL et al 2002 Bacteria-triggered $CD4^+$ T regulatory cells suppress *Helicobacter hepaticus*-induced colitis. J Exp Med 196:505–515

Le Goff L, Lamb TJ, Graham AL, Harcus Y, Allen JE 2002 IL-4 is required to prevent filarial nematode development in resistant but not susceptible strains of mice. Int J Parasitol 32: 1277–1284

Mahanty S, Mollis SN, Ravichandran M et al 1996 High levels of spontaneous and parasite antigen-driven interleukin-10 production are associated with antigen-specific hyporesponsiveness in human lymphatic filariasis. J Infect Dis 173:769–773

Maizels RM 2005 Infections and allergy–helminths, hygiene and host immune regulation. Curr Opin Immunol 17:656–661

Maizels RM, Yazdanbakhsh M 2003 Regulation of the immune response by helminth parasites: cellular and molecular mechanisms. Nat Rev Immunol 3:733–743

Maizels RM, Gomez-Escobar N, Gregory WF, Murray J, Zang X 2001 Immune evasion genes from filarial nematodes. Int J Parasitol 31:889–898

McGuirk P, McCann C, Mills KHG 2002 Pathogen-specific T regulatory 1 cells induced in the respiratory tract by a bacterial molecule that stimulates interleukin 10 production by dendritic cells: a novel strategy for evasion of protective T helper type 1 responses by *Bordetella pertussis*. J Exp Med 195:221–231

Mellor AL, Chandler P, Baban B et al 2004 Specific subsets of murine dendritic cells acquire potent T cell regulatory functions following CTLA4-mediated induction of indoleamine 2,3 dioxygenase. Int Immunol 16:1391–1401

Pearce EJ, Kane CM, Sun J, Taylor JJ, McKee AS, Cervi L 2004 Th2 response polarization during infection with the helminth parasite *Schistosoma mansoni*. Immunol Rev 201:117–126

Peisong G, Mao X-Q, Enomoto T et al 2004 An asthma associated genetic variant of STAT6 predicts low burden of *Ascaris* worm infection. Genes Immun 5:58–62

Platts-Mills T, Vaughan J, Squillace S, Woodfolk J, Sporik R 2001 Sensitisation, asthma, and a modified Th2 response in children exposed to cat allergen: a population-based cross-sectional study. Lancet 357:752–756

Powrie F 1995 T cells in inflammatory bowel disease: protective and pathogenic roles. Immunity 3:171–174

Sakaguchi S 2004 Naturally arising $CD4^+$ regulatory T cells for immunologic self-tolerance and negative control of immune responses. Annu Rev Immunol 22:531–562

Sartono E, Kruize YCM, Kurniawan-Atmadja A, Maizels RM, Yazdanbakhsh M 1997 Depression of antigen-specific interleukin-5 and interferon-γ responses in human lymphatic filariasis as a function of clinical status and age. J Infect Dis 175:1276–1280

Satoguina J, Mempel M, Larbi J et al 2002 Antigen-specific T regulatory-1 cells are associated with immunosuppression in a chronic helminth infection (onchocerciasis). Microbes Infect 4:1291–1300

Shevach EM, McHugh RS, Piccirillo CA, Thornton AM 2001 Control of T-cell activation by $CD4^+CD25^+$ suppressor T cells. Immunol Rev 182:58–67

Shimizu J, Yamazaki S, Takahashi T, Ishida Y, Sakaguchi S 2002 Stimulation of $CD25^+CD4^+$ regulatory T cells through GITR breaks immunological self-tolerance. Nat Immunol 3: 135–142
Steel C, Nutman TB 2003 CTLA-4 in filarial infections: implications for a role in diminished T cell reactivity. J Immunol 170:1930–1938
Suffia IJ, Reckling SK, Piccirillo CA, Goldszmid RS, Belkaid Y 2006 Infected site-restricted $Foxp3^+$ natural regulatory T cells are specific for microbial antigens. J Exp Med 203: 777–788
Suvas S, Kumaraguru U, Pack CD, Lee S, Rouse BT 2003 $CD4^+CD25^+$ T cells regulate virus-specific primary and memory $CD8^+$ T cell responses. J Exp Med 198:889–901
Symula DJ, Frazer KA, Ueda Y et al 1999 Functional screening of an asthma QTL in YAC transgenic mice. Nat Genet 23:241–244
Taylor M, Le Goff L, Harris A, Malone E, Allen JE, Maizels RM 2005 Removal of regulatory T cell activity reverses hyporesponsiveness and leads to filarial parasite clearance in vivo. J Immunol 174:4924–4933
Ueda H, Howson JM, Esposito L et al 2003 Association of the T-cell regulatory gene CTLA4 with susceptibility to autoimmune disease. Nature 423:506–511
van den Biggelaar A, van Ree R, Roderigues LC et al 2000 Decreased atopy in children infected with *Schistosoma haematobium* : a role for parasite-induced interleukin-10. Lancet 356:1723–1727
van den Biggelaar AH, Borrmann S, Kremsner P, Yazdanbakhsh M 2002 Immune responses induced by repeated treatment do not result in protective immunity to *Schistosoma haematobium*: interleukin (IL)-5 and IL-10 responses. J Infect Dis 186:1474–1482
van der Pouw Kraan TC, van Veen A, Boeije LC et al 1999 An IL-13 promoter polymorphism associated with increased risk of allergic asthma. Genes Immun 1:61–65
Vella A, Cooper JD, Lowe CE et al 2005 Localization of a type 1 diabetes locus in the IL2RA/CD25 region by use of tag single-nucleotide polymorphisms. Am J Hum Genet 76:773–779
Wilson MS, Taylor M, Balic A, Finney CAM, Lamb JR, Maizels RM 2005 Suppression of allergic airway inflammation by helminth-induced regulatory T cells. J Exp Med 202: 1199–1212
Zang X, Maizels RM 2001 Serine proteinase inhibitors from nematodes and the arms race between host and pathogen. Trends Biochem Sci 26:191–197
Zang XX, Taylor P, Meyer D et al 2002 Homologues of human macrophage migration inhibitory factor from a parasitic nematode: gene cloning, protein activity and crystal structure. J Biol Chem 277:44261–44267

DISCUSSION

Morahan: How are the worms mediating this trick? Are they producing a factor that is stimulating the T_{regs}?

Maizels: Yes, we have to hypothesize that they make molecular products that can drive that differentiation. We have several candidates. One is a TGFβ homologue. These multicellular animals have multiple TGFβ family members, most of which play a developmental role. We have found one that is secreted (Gomez-Escobar et al 2000). It would be a good trick for the parasite to take advantage of the immunological potential of these products. For example, TGFβ can induce FoxP3 expression in naïve T cells (Chen et al 2003). We also have other candidates from novel gene families that may be mediating this type of development.

Malissen: What is the antigen specificity of these T_{regs}?

Maizels: The hot question is whether these T_{regs} are parasite antigen specific or not. There is some evidence that they are. If we co-culture a T cell population with parasite antigen, we can show an expansion in the TGFβ-expressing cell population. On the other hand if we deplete the pre-existing $CD25^+$ cells before the mouse has seen the parasite, then the infection is abbreviated. That result argues that natural T_{regs} also make a contribution.

Cyster: Is the expansion of the T_{regs} population restricted to the mucosal compartment, and are they homing to places like the lung?

Maizels: We have looked in the spleens of mice carrying an intestinal infection and find the same expansion, but to a less marked degree. In terms of migration to the lung, we have looked in transfer models using allotype-marked donor cells. The transferred cells distribute themselves throughout the body, including the lung. They seem to be far ranging.

Turner: You have showed the immunosuppression with allergy. What about autoimmunity?

Maizels: We have started looking at this with the EAE model. There is an effect, but it is not as marked. We haven't yet shown that this is the CD25 population. I would like to do these experiments more extensively because it is a key question whether the same cell can suppress both the Th1 and Th2.

Turner: Related to the specificity question, can you take any old T_{reg} and then show this suppression of allergy?

Maizels: So would purified T_{regs} from a naïve animal and have the same effect? I imagine if you give enough it would work. This is how the original type of Fiona Powrie experiments were done. It may be a numbers game, or it may be that it is the number of activated T_{regs} is important. If you had enough activated cells from a naïve source this could have the same effect.

Turner: It would be interesting to know if there is something specific about infection that is activating the T_{regs}.

Vinuesa: You mentioned that in the *Schistosoma*-infected children, IL10 was protective against the atopy. But in this model the IL10 comes from the Th2 cells. Does IL10 play a protective role in this system?

Maizels: In the mouse system we have found that IL10 is not required for down-regulation by helminths. This can't be extrapolated to prejudge the human situation. Different cells make IL10 in humans and mice, and there may also be a time phenomenon: the mouse models are abbreviated, while IL10 may have a role in the long-term homeostatic context rather than in a more compressed experimental time frame.

Britton: These effector mechanisms may have different effects in different end organs. One example is the suggestion that BCG vaccination is associated with reduced IgE responses and reduced asthma. We examined the rates of allergy and asthma in children in Sydney with and without BCG vaccination at birth. BCG vaccination had no effect on house dust mouse specific or total IgE responses, but was associated with a reduction in asthma in an environment with high exposure to allergens (Marks et al 2003). Therefore BCG influenced the expression of allergic responses only in the lung. This highlights the fact that the simple Th1/Th2 dichotomy is not relevant here.

Hume: You didn't say 'macrophage' once. With respect to parasites, there are a lot of people working on direct effects of parasite extracts on macrophages and DCs. Do you think the T_{reg} polarization is a read out of the nature of antigen presentation? There is an interesting recent observation that the F480 knockout doesn't make T_{regs}. This suggests that there is a molecule that is utterly macrophage restricted that in some measure determines.

Maizels: Yes, we must have a hypothesis that says that the initial antigen-presenting population is selecting and driving differentiation of the antigen-specific T cells. But there is also the possibility that parasites have learned to expand the pre-existing population of self-reactive T cells. This may not need an antigen processing cell as a mediator.

Lam: Is there a role for B cells in control of these parasites? In human, we have a population of B1 cells which are known to secrete IL10.

Maizels: Yes. We found that the two populations that had surface TGFβ were the T_{reg} and the B cell. In the control of EAE the B cell population was more active than the T cell population. I can't yet put these into a sequential order. The presence of the regulatory expanded network may spill over and alter the phenotype of the B cell population. Again, there is nothing that tells us that this is IL10 dependent.

Cook: I'd like to return to the point about the nature of the end organ inflammation. You have shown that the effect is to some extent independent of the IgE levels. There is obviously some inflammation there. Do you know if there is a direct effect on airway hyperreactivity?

Maizels: We haven't yet tested this, but we intend to use non-invasive plethysmography for exactly this purpose; this also has the great attraction of enabling us to do sequential measurements on the same animal. This is more relevant to the concept that these organisms or their products may have some therapeutic value.

Cook: So at the moment the idea is that there is something that is inhibiting the recruitment of the effector population. This is then acting on the target organ to have the desired physiological effect, rather than there being a direct effect of the T_{regs} on the end organ. Is that right?

Maizels: I can't distinguish among these options at the moment. The first sounds more plausible, but we can't exclude a direct effect.

Cyster: Another part of the hygiene hypothesis is that the commensal flora may also be changing. Do you think that changes in the commensal flora could play out in a similar way?

Maizels: Definitely. There are some tantalizing data showing that young children with asthmatic symptoms have different commensal species from children who don't. There are also data showing that asthmatic symptoms in children are more common in those who have had antibiotics in the first year of life, possible due to perturbation of the commensal flora.

References

Chen W, Jin W, Hardegen N et al 2003 Conversion of peripheral CD4+CD25- naive T cells to CD4+CD25+ regulatory T cells by TGF-β induction of transcription factor Foxp3. J Exp Med 198:1875–1886

Gomez-Escobar N, Gregory WF, Maizels RM 2000 Identification of Bm-tgh-2, a filarial nematode homolog of C. elegans daf-7 and human TGF-β, expressed in microfilarial and adult stages of Brugia malayi. Infect Immun 68:6402–6410

Marks GB, Ng K, Zhou J et al 2003 The effect of neonatal BCG vaccination on atopy and asthma at age 7 to 14 years: an historical cohort study in a community with a very low prevalence of tuberculosis infection and a high prevalence of atopic disease. J Allergy Clin Immunol 111:541–549

Closing remarks

Chris Goodnow

John Curtin School of Medical Research, The Australian National University, Canberra, ACT 2061, Australia

I'd like to sum up some of the themes we have discussed at this meeting. It is hard to summarize all the topics we touched on, but I'd like to address a few of the recurring themes and challenges.

We talked a lot about the large number of sequences that encode the building blocks for the immune response. We have heard about highly complex networks of interacting genes. To what extent are these going to prove hopelessly complicated spaghetti diagrams versus elegant bow-tie convergence points? At this stage we don't have good clear examples of how complex gene networks work and how these networks might be teased out.

Another theme is the spectrum of kinds of genetic variation in terms of the relationship between genetic variation and phenotypes. These range from simple Mendelian traits, to genes of large effect masked by lots of modifiers thrown in, right through to QTLs that contribute small amounts to phenotypic variance.

Philippe Gros brought together an issue we have seen in many of the papers: on the one hand the simple single-gene effects are the easiest fruit to pick off the tree, but on the other hand they aren't the only ones. If we want to get to grips with how the code really works, we need to deal with the QTLs. We heard a number of examples of success in identifying QTLs in tough diseases such as MS and SLE.

One of the things that emerges is the relationship between immune responses and different kinds of genetic architectures. We had examples of simple types of genetic variation such as single nucleotide substitutions that change amino acids, right through to much more complex genetic changes, including duplications of genetic regions (such as TLR7) and complex copy number changes in loci between strains of mice. Each of these will be major sources of variation in human and animal populations. As we also heard, the same kinds of genetic variation are going on in pathogens, which the latter are using to tune their side of the host/pathogen equation.

Another theme we keep circling around is this whole issue of where innate immunity stops and adaptive immunity starts. When we say something is a problem with innate immunity, is it really? We talked abut TLR7 gene variants being a difference in innate immunity, but as Ward Wakeland has shown, this innate immune sensor is having its effect intrinsically in a B cell, which is an adaptive immune cell. These are intertwined, to such an extent that it makes less sense to keep drawing a distinction between innate and adaptive immunity.

This brings me to the theme of kinetics. One of the biggest problems we face is kinetics. We talked about how the whole course of the disease may be determined by subtle genetic differences that act in the first days after infection, when there are just a few organisms, yet their effects are manifest perhaps months later. For example, we talked about effects of the IRF8 mutation, where this might be reflecting something that happens early in the immune response, but it might be equally reflecting a loss of a particular dendritic cell subset that causes exhaustion of the T cells much later on.

Perhaps the biggest problem we have seen articulated is that for some important immune phenomena we need experiments with a long time scale.

Experiments where we have to follow a mouse for 200 days or a human cohort for 20 years rapidly fall out of the reach of the time scales of the grants that fund our labs. This is an interesting issue: it is possible to tease out these complex gene–environment–pathogen interactions, but the success stories have often involved tenacity and people sticking with a problem for a decade or more. This is an important message. We will be successful, but it will require patience. It would be easy to say that the problems are too hard to solve, but what this meeting has given us is inspiring examples of how tenacity, new tools, time and insightful approaches have led to tremendous progress at cracking the immune system's genome–phenome code. At the same time, we are only at the tip of the iceberg. As the genetics unfolds, it will be interesting to see where this field has got to in 10 years time. Ten years ago, we didn't know about TLRs or FoxP3, for example. We are clearly on the upswing of understanding the genomic control of the immune response, not the end-game. I thank you all for your participation.

Contributor Index

*Non-participating co-authors are indicated by asterisks. Entries in **bold** indicate papers; other entries refer to discussion contributions.*

Subject Index

I

S

T

U

V

W

X

Y

Z